B292
Management accounting

Business School

Unit 1

The role and context of management accounting

Written by Mike Lucas

Module Team

Dr Mike Lucas, *B292 Chair & Author*

Professor Jane Frecknall-Hughes, *Professional Certificate in Accounting Chair & Author*

Elizabeth Porter, *Regional Manager & Author*

Jonathan Winship, *Author*

Stuart Munro, *Author*

Dr Vira Krakhmal, *Author*

Dr Pauline Gleadle, *Author*

Dr Jane Hughes, *Contributor*

Sam Cooper, *Programme Coordinator*

Emir Forken, *Programme Manager*

Dr Lesley Messer, *Head of Curriculum Operations*

Funmi Mapelujo, *Qualifications Manager*

Kelly Dobbs, *Curriculum Assistant*

External Assessor

Professor Stuart Turley, Manchester Business School

Critical Readers

Richard Davies

Dr Jane Hughes

Developmental Testers

Dr Teodora Burnand

Sam Cooper

Diane Jamieson

Sue Winship

Nicole Wright

Production Team

Simon Ashby, *Media Developer*

Martin Brazier, *Media Developer*

Anne Brown, *Media Assistant*

Siggy Martin, *Print Buyer*

Lee Johnson, *Media Project Manager*

Vicky Eves, *Media Developer*

Diane Hopwood, *Rights Assistant*

Kelvin Street, *Library*

Keith Wakeman, *Service Administrator*

The Module Team wishes to acknowledge use of some materials from B680 *The Certificate in Accounting.*

This publication forms part of the Open University module B292 *Management accounting*. Details of this and other Open University modules can be obtained from the Student Registration and Enquiry Service, The Open University, PO Box 197, Milton Keynes MK7 6BJ, United Kingdom (tel. +44 (0)845 300 60 90; email general-enquiries@open.ac.uk).

Alternatively, you may visit the Open University website at www.open.ac.uk where you can learn more about the wide range of modules and packs offered at all levels by The Open University.

To purchase a selection of Open University materials visit www.ouw.co.uk, or contact Open University Worldwide, Walton Hall, Milton Keynes MK7 6AA, United Kingdom for a brochure (tel. +44 (0)1908 858793; fax +44 (0)1908 858787; email ouw-customer-services@open.ac.uk).

The Open University
Walton Hall
Milton Keynes
MK7 6AA

First published 2011. Second edition 2012.

Edited and designed by The Open University.

Typeset in India by OKS Prepress Services, Chennai.

Printed in the United Kingdom by Henry Ling Limited, at the Dorset Press, Dorchester, DT1 1HD.

ISBN 978 1780 0 7380 4

2.1

Contents

Introduction

Welcome to the first unit of B292 *Management accounting*.

In Unit 1, there are four sessions, corresponding to each of the unit's main aims. This unit introduces you to the nature of management and the role of management accounting in the management process. This knowledge is essential for understanding the subsequent six units, which together make up the B292 *Management accounting* module. Management accounting is concerned with providing information and analysis to managers to help them plan, evaluate and control activities, in order to achieve an organisation's objectives. Whereas financial accounting is concerned with reporting on the past financial performance of an organisation, management accounting is essentially concerned with improving its future performance. The nature and role of management accounting in organisations will be discussed in detail in Session 2 of this unit. In order to understand the concepts and principles of management accounting, however, it is necessary first to have some appreciation of what managers do! This, in turn, requires an understanding of the organisations in which managers work – and of the external environment in which these organisations exist and operate.

Session 1 of this unit, therefore, looks at the nature of organisations, specifically their objectives and structure. Organisational objectives and structure are key elements of organisations and they determine management functions and responsibilities within the organisation. Session 1 also considers the main environmental factors (economic, social, political, legal and technological) that impact on organisational behaviour.

Session 2 then looks in greater depth at the management processes of **planning, control** and **decision making**, as they operate in organisations. The principal role of management accountants is to provide information and analysis to help managers with their planning, control and decision making activities. This role will be explored in this session.

Session 3 considers the role of information and **information systems** in organisations, an understanding of which is important for the design and operation of effective **management accounting systems (MAS)**.

Session 4 then considers the opportunities provided by **information technology** for the functioning of management accounting systems in an organisation. To function effectively and efficiently, management accountants must have a good knowledge and understanding of such technology.

By the end of Unit 1, you should have a clear idea of the nature of organisations, the process of management and the role of information (including accounting information) in managing organisations. You should also have an appreciation of the opportunities presented by information technology for providing management information.

Learning aims and outcomes of Unit 1

Upon completion of Unit 1 you are expected to be able to understand and explain:

1 the nature and objectives of organisations
2 the nature of management and the role of management accounting in the management process
3 the role of information and information systems in organisations
4 the opportunities offered by information technology for managing organisations.

SESSION 1 The nature of organisations: objectives, structure, management functions and responsibilities

Introduction

Upon completion of Session 1, you are expected to be able to:

- understand the nature and purpose of different types of organisations (commercial, voluntary, public sector and so on)
- describe the different ways in which organisations may be structured
- understand basic concepts of organisational structure (span of control, scalar chain, separation of direction and management, tall and flat organisations)
- describe the main departments or functions of a business organisation
- explain the advantages and disadvantages of centralised and decentralised organisations
- describe the different types of responsibility centres and the information needs of their managers
- understand the impact of environmental factors (political, legal, economic, social and technological) on organisations.

In this session you will look at the nature of organisations: their objectives, structure and how they are managed. Knowledge and understanding of these topics is necessary in order to understand the role that management accounting plays in organisations – which is introduced in Session 2 of this unit. Management accountants are an integral part of the management team of an organisation. In order to provide accounting information relevant to managers' needs, it will be necessary for management accountants to understand the nature and functioning of their organisation, including the operations of the various functional areas of the organisation.

1.1 What is an organisation?

An 'organisation' is a group of individuals working together to achieve one or more objectives. Although organisations have been defined differently by different theorists, virtually all definitions refer to five common features:

1. they are composed of individuals and groups of individuals
2. they are oriented towards achieving collective goals
3. they consist of different functions
4. the functions need to be coordinated
5. they exist independently of individual members who may come and go.

1.1.1 Mintzberg's five components of organisation

Mintzberg (1979, p. 24) suggested that all organisations consist of five components, as shown in Figure 1.

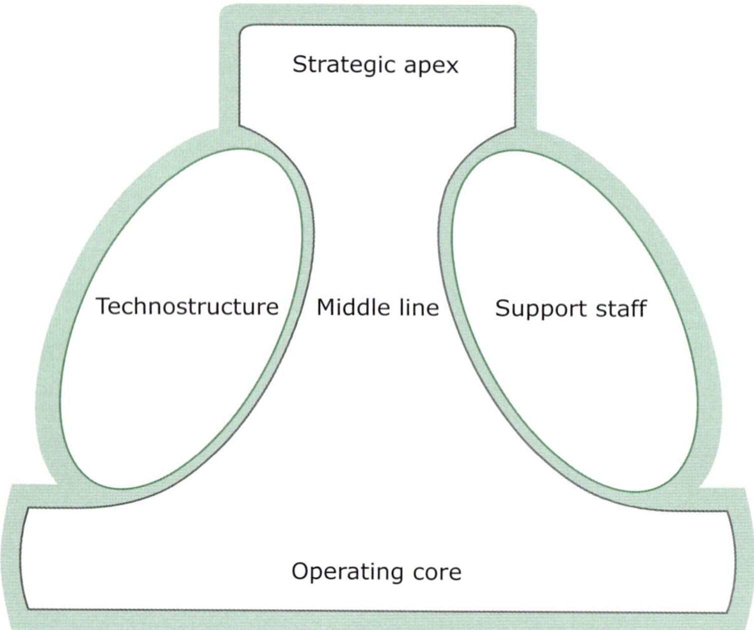

Figure 1 Mintzberg's five parts of the organisation

At the top of the organisation is a **Strategic apex** the purpose of which is to ensure the organisation follows its mission and manages its relationship with its environment. The individuals comprising the apex, for example, the Chief Executive Officer (CEO), are responsible to owners, government agencies, unions, communities and so on.

Below the apex is the **Middle line**, a group of managers who are concerned with converting the objectives and broad plans of the Strategic apex into operational plans that can be carried out by the workers.

As organisations grow and become more complex, they usually develop a separate group of people who are concerned with the best way of doing a job, specifying output criteria (e.g., quality standards) and ensuring that personnel have appropriate skills (e.g., by organising training programmes). This group of analysts is referred to by Mintzberg as the **Technostructure**. The organisation also adds other administrative functions that provide services to itself, for example legal advice, public relations, mailroom, cafeteria and so on. These are the **Support staff**.

Finally, at the bottom of the organisation, is the **Operating core**. These are the people who do the basic work of producing the products or delivering the services.

Mintzberg's generic organisational model also illustrates an important principle of organisation structure: the separation of direction and management, whereby those people who decide the mission and general direction of the organisation are different (other than in a very small organisation) from those who handle the implementation of plans and subsequent controlling of operations to ensure that objectives are met. Senior managers (the Strategic apex) will establish long-term organisational objectives and policies through which goals are to be achieved. Middle managers (the Middle line) will be responsible for translating the necessarily broad and general strategic plans into detailed action plans, specifying managerial responsibilities for particular tasks and how resources are to be allocated. These middle managers will also be responsible for monitoring activities and taking action to ensure that resources are being used **efficiently** and **effectively** to achieve organisational objectives.

Efficiency refers to the relationship between inputs and outputs. An activity or process is efficient, if it produces a given output with the minimum of inputs necessary. Effectiveness refers to the extent to which goals/objectives are actually achieved.

Other important principles of organisational structure are discussed later in Session 1 (see Sections 1.2 and 1.3).

1.1.2 Why do organisations exist?

Organisations exist because groups of people working together can achieve more than the sum of the achievements which the individuals in the organisation could produce when working separately. For example, one person might struggle all day to carry a piano upstairs, whereas a team of four people, each taking one corner, may need to put in much less than a quarter of the effort of one person to complete the task (Coates et al., 1996, p. 19). Although such cooperation is beneficial, if individuals pull in different directions, the result is counter-productive. Thus coordination is necessary and this is a fundamental role of management, as will be discussed in a later section of this session.

It can also be argued that organisations exist as a result of the impact of **transaction costs**, because they can arrange transactions between their different parts at a lower total cost than that available in the open market. In crude terms, it may be cheaper to make or do something 'in house' because this cuts out the time consuming process of negotiating terms – and renegotiating them every time your requirements change. Before the Industrial Age, it was common for artisans/craftsmen to work individually from home, producing various products, which merchants would purchase from these

individuals and sell to consumers. All transactions were conducted through market exchange without the need for formal organisations. As economic development gained pace, however, it became clear that it was more efficient to organise production internally within a firm rather than undertake each transaction externally through the market. The latter process involved substantial costs in terms of time that had to be spent investigating suppliers, detailing specifications, negotiating/renegotiating contracts, checking that agreed terms had been met and so on.

1.1.3 What types of organisation are there?

Organisations can be classified in different ways. One way is according to their over-arching purpose, or primary objective. Broadly, organisations may be classified as 'for-profit' (i.e., commercial) or 'not-for-profit' entities. **'For-profit' (commercial) organisations** may have several different objectives. For a very long time, it was generally accepted that maximising the wealth of the owners and continuing in existence were the primary objectives of profit seeking organisations. However, as organisations also aim, for example, to provide goods and services to customers and employment to employees, it is perhaps more reasonable to suggest that increasing, rather than maximising the wealth of owners, is a more fitting objective. **'Not-for–profit' organisations** comprise a large variety of organisations including charities, clubs, **cooperative firms/social enterprises** and public sector organisations. Public sector organisations are owned, funded and run by central or local government. They include:

- public hospitals
- the armed forces (military)
- most schools and universities
- government departments.

These organisations exist to provide services which, for various reasons, it is considered impractical or undesirable for the commercial sector to provide.

Whereas commercial organisations, charities and social enterprises must generate sufficient funds from their activities to sustain themselves on a continuing basis, public sector organisations are funded by government. Nevertheless, constraints on government expenditure mean that resources are limited. Consequently, economic scarcity requires that virtually all organisations be run effectively and efficiently. As a result, many of the management principles employed by the commercial sector are also employed in the not-for-profit sector, requiring extensive use of management accounting in all sectors.

Maximisation of shareholder value has long been the publicly stated objective of most business enterprises. It is likely, however, partly as a result of the global financial crises that began in 2008, that the publicly stated objectives will be expanded to embrace more stakeholders, such as employees and the local community.

A traditional view of differences between sectors is illustrated in Figure 2. However, these distinctions are becoming blurred, as indicated by the overlapping circles. Commercial organisations are increasingly pursuing **social responsibility objectives**, while not-for-profit organisations are increasingly adopting commercial criteria to ensure the sound financial management of scarce resources.

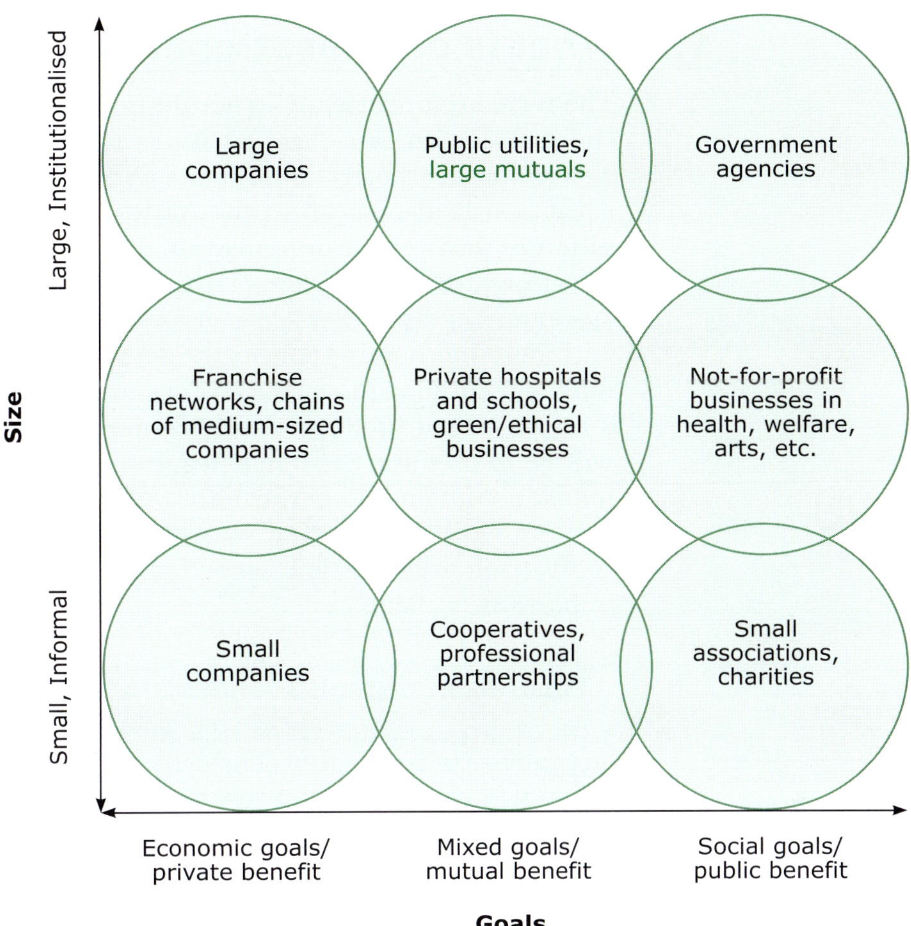

Figure 2 Organisations highlighting differences between sectors

Activity 1.1

(Remember, you should never look at the answer that follows before attempting an activity.)

- Think of an organisation that you know well. Which factors (profit, accountability, commitment) exert pressure on it or influence its objectives?

Mark on Figure 2 the sector where you think your organisation 'sits'.

Think about the influences that these factors have on the organisation and its work.

Feedback

You may have decided that your organisation sits clearly inside one of the circles or, more likely, you will have sited it in an intermediate position – operating for commercial purposes, but anxious to satisfy other criteria such as accountability to a wider public. For example, there has been a rise in the number of investment companies which use not only financial but also ethical criteria to decide in which organisations to invest. Such criteria could include the environmental impact of the production processes used by the companies in question, or the working conditions and wages of their employees. Another example of the mixture of the pressures that can affect organisations could be taken from the social care sector: a organisation in this sector might be keen to provide an excellent service to local people, but it could also have to operate on commercial principles in managing costs in order to compete with more cost effective providers.

1.2 What is organisational structure?

The term organisational structure refers to the relationships between the various functions and positions in an organisation. Structure determines authority and responsibility for particular tasks/activities. It also specifies the routes of communication between different parts of the organisation. Organisational structure therefore has important implications for the design of management accounting systems. For example, some organisations are highly decentralised, with decision making authority delegated to relatively junior managers at lower levels in the organisational hierarchy. In this case, a major role of the organisation's management accounting system will be to monitor the outcomes and provide feedback to senior managers about the performance of those who have the decision making authority. Such a role will not be necessary in a highly centralised organisation, where senior managers make all the important decisions.

Theories of organisational structure

With the emergence of large industrial enterprises in the nineteenth century, management theorists began to consider how organisations should be designed and managed. As you will read later in this session, early organisational theorists such as Henri Fayol (1949) attempted to derive universal prescriptions for the optimal design of organisations. More recently, theorists have emphasised the contingent nature of optimal organisational design – depending on variables such as size, production technology, degree of stability in the organisation's business environment, nature of competition in the industry and so on. These factors are assumed to influence, for example, whether organisations are 'tall' or 'flat', centralised or decentralised in terms of decision making and so on. (These terms are discussed later.)

Ultimately, organisational structure is a means of influencing and controlling the behaviour of the individuals who work in the organisation. Structure is used to assign authority and responsibility to individuals and hold them accountable for the achievement of specific tasks or objectives (Emmanuel et al., 1990, p. 38). Management accounting is an important part of this process as it provides managers with the information to carry out the various activities for which they are responsible. It also measures and monitors their performance to ensure that the organisation achieves its objectives.

A good example of a management accounting technique employed widely for organisational planning and control purposes is budgeting.

The sort of information managers need to undertake various activities and the way their performance is measured/monitored will depend on the way in which the organisation is structured.

1.2.1 Principles of organisational structure

Certain principles are basic to the operation of any organisation:

Specialisation

The work of the organisation is divided up into separate activities or tasks and particular individuals concentrate on specific tasks or activities. This enables the application of specialised knowledge and so improves organisational efficiency and effectiveness.

Coordination

If an organisation's activities are to be separated into different areas or operations, it will be necessary to ensure that the various actions are coordinated, that is, consistent with each other and working towards the same organisational objectives. This is a very important task of management. The management hierarchy or 'chain of command' facilitates the coordination of various departments and their activities.

Management principles of the hierarchy of authority

Management theorists (notably Henri Fayol, 1949) have, over the years, developed several principles relating to the hierarchy of authority for coordinating activities. Some of the most important are:

- Unity of Command. Every person should receive orders and be accountable to one and only one superior. If people receive orders from more than one superior, conflict and confusion may well result.
- The Scalar Chain. There should be a clear line of authority from top to bottom, linking all managers at all levels.
- The Responsibility and Authority Principle. If an organisational member is allocated responsibility, then that person should also be given the necessary authority to carry out the tasks necessary – including the right to ask other people to undertake particular tasks. A manager should not be given responsibility without the necessary authority, but conversely delegation of authority implies responsibility and the need for accountability.

Span of control refers to the number of subordinates directly reporting to a manager or supervisor.

- Span of Control. There is a limit to the number of activities or people that can be supervised effectively by one person. What constitutes an effective span of control will be determined by a number of factors, including:
  - the similarity of tasks/functions undertaken (the more similar, the greater the potential effective span of control)
  - the proximity of the tasks to each other and to the supervisor (the closer the proximity, the greater the potential effective span of control)
  - the complexity of the tasks (the more complex, the smaller the potential effective span of control)
  - the direction and control needed by subordinates (the more direction and control needed, the smaller the potential effective span of control).

Activity 1.2

Consider the organisation where you (or a close friend or relative) work or have worked.

- What is the span of control of your immediate manager?
- What is the span of control of her or his manager?
- Do you think they are appropriate, bearing in mind the ability of either person to monitor what is going on in the organisation?

Feedback

The appropriateness of a span of control may depend on the extent of the delegation that can be exercised by managers (i.e., entrusting the responsibility for tasks to someone else) and also on reporting mechanisms within the organisation.

Many organisations have introduced systems of regular meetings between managers and staff at which SMART objectives are set and monitored. SMART objectives are:

- **S**pecific
- **M**easurable
- **A**chievable
- **R**ealistic
- **T**ime-bound (i.e., have a defined time scale associated with their achievement).

If objectives are SMART, managers should be able to tell whether or not they have been achieved. The appropriateness of the span of control will then be related to the number of people who can realistically be monitored in this way and the frequency with which monitoring takes place.

1.2.2 Tall versus flat organisations

Where there is a large number of levels in the management hierarchy, the organisation is said to be 'tall'. This will tend to result in narrow spans of control. Where there is a small number of levels in the hierarchy, the organisation is said to be 'flat'. Flat organisations will tend to have wide spans of control. Figure 3 shows a comparison of tall and flat organisation structures.

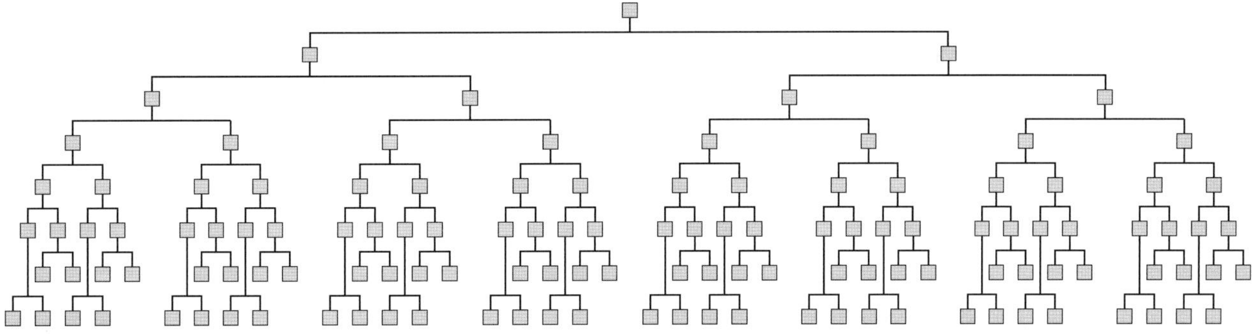

A tall organisation

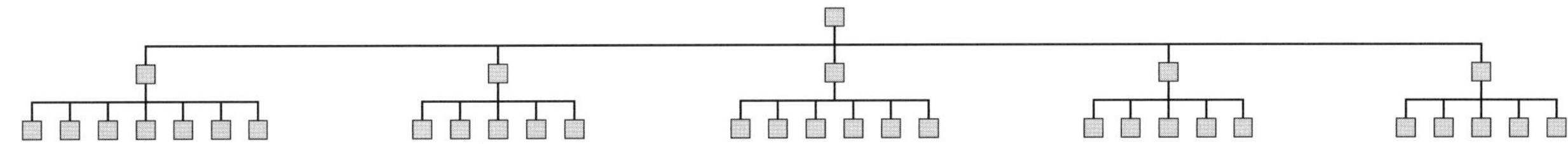

A flat organisation

Figure 3 Tall and flat organisations

In recent years, there has been a trend towards **delayering**, whereby tall organisations have tended to become flat organisations by the removal of various levels in the hierarchy. This has been facilitated by:

(a) Information technology, which has reduced the need for many middle management jobs, which were largely concerned with processing information to facilitate control within the organisational hierarchy.

(b) The management philosophy of **empowerment**, whereby people at lower levels have been delegated authority to take actions and make decisions which would previously have been the domain of middle managers.

Changes in organisation structures have led to changes in approach to management and vice versa.

The advantages of delayering are:

(a) A significant reduction in costs as middle managers' salary costs are removed.

(b) Increased motivation as people at lower levels are given power/ discretion to make decisions. Improved performance is likely to be a consequence of increased motivation.

(c) Improved, faster communication between senior management and operational levels – increasingly important in a faster changing, more uncertain and increasingly competitive external environment.

There are also disadvantages to delayering, the principal one being a possible loss of control. Middle managers are often necessary to translate the inevitably broad and general plans of senior management into operational plans and actions that can be implemented. Senior managers may have only a hazy understanding of what is going on at the operational level and much is thereby entrusted to relatively junior people (Coates et al., 1996, p. 116).

Activity 1.3

- How will a tall organisational structure, in contrast with a flat one, impact on the span of control and the speed of information flows through the organisation?
- What do you think are the advantages and disadvantages of flat organisations?

Feedback

In tall organisations, managers have smaller spans of control (i.e., fewer people reporting directly to them). This reduces the number of people they have to manage, but means it takes longer for information to travel through the layers of the organisation.

In flat organisations, the opposite is true: communication can be quicker because of fewer layers, but the spans of control are larger. In recent years, tall organisations have tended to be associated with large bureaucracies. Communication will be formal and middle managers may be in a position to use information as a device to retain control. On the other hand, smaller spans of control may mean that managers have more time to manage. Furthermore, there are more explicit career paths, with opportunities for promotion.

Flat organisations have developed because there is a belief that communication is impaired by additional levels of management. Flatter organisations are thought to be able to react to change more quickly. Managers may be forced to delegate if their span of control is enlarged. This can be motivating for those to whom work is delegated. On the other hand, career paths are less explicit: employees may have to look sideways or even outside the organisation for career development opportunities.

1.3 Forms of organisational structure

It is no easy task to design and develop a structure. Many organisations continuously debate whether to structure around products, geography, common tasks or information.

It is possible to distinguish six important forms of organisational structure:

1 functional structure
2 product or service structure
3 geographical structure
4 matrix structure
5 project team
6 hybrid structure.

Each of these forms is now explained in turn.

1.3.1 Functional structure

As you can see from Figure 4, in a functional structure people are grouped according to the type of job they do.

A functional structure

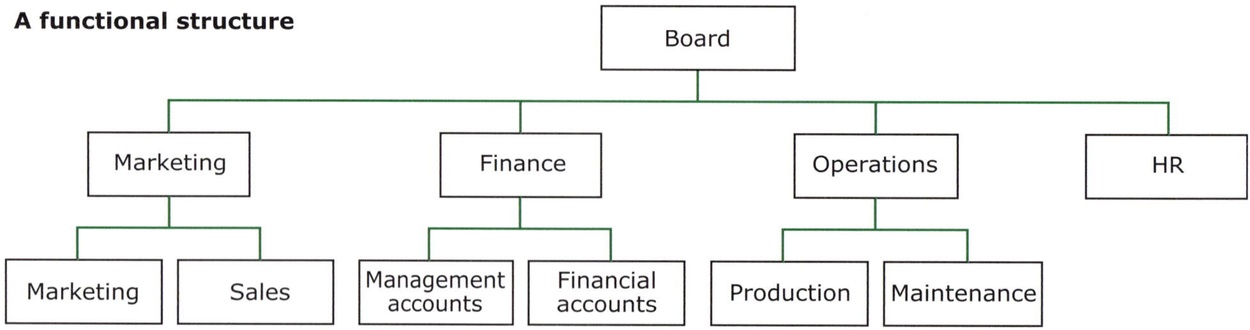

Figure 4 A functional structure

This structure may be appropriate when people within functional departments need to communicate regularly with each other. For example, in a marketing department, the marketing director will coordinate the activities of marketing specialists in fields such as promotion, advertising, product design, market research, and packaging. Although there is a need for communication with all the other parts of the company, the bulk of the information exchange and communication is likely to be within the functional areas, so it makes sense to group these people together.

Functional structures, however, can have disadvantages.

- Career paths tend to develop through functions and this can reduce managers' awareness of other issues facing the organisation. The organisation will not develop many generalists this way (e.g., people who know about marketing and operations or finance and HR). This will expose the organisation to significant risks, as managers will not have an overview of its operations.
- Staff may work for the benefit of their department and not the organisation as a whole.
- Many members of staff may never meet an external customer and may not therefore have a customer service orientation.

1.3.2 Product or service structure

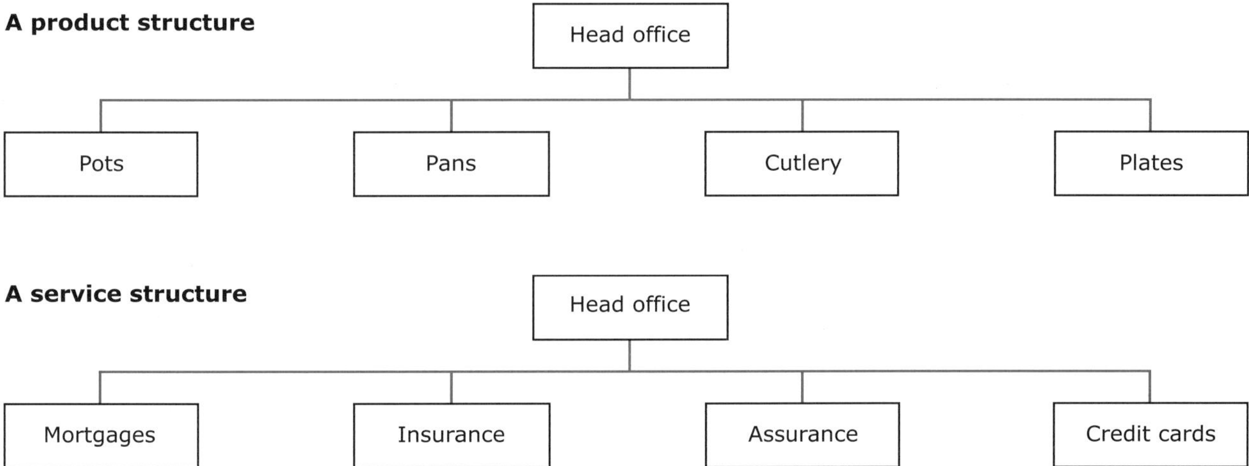

Figure 5 Product and service structures

At first glance this looks similar to the functional structure, but here staff members are grouped together on product or service lines. This very often happens in larger organisations. For example, a big accountancy firm may group staff by the industry served or a public education department may group staff around different areas of work (e.g., pre-school children, primary, secondary and special needs). Each product or service group will have its own production and service people and also its own accounting and personnel staff. A product or service structure can be more responsive to customer needs and better at motivating staff. However, there is a danger of creating independent units, which can be difficult to manage as they assume the attitude, *'we know what the customer wants, so stop interfering!*'. It can also mean that professional expertise becomes fragmented. For example, if each product line has its own small accounting team (perhaps just one person) this may reduce career opportunities for specialists.

1.3.3 Geographical structure

A geographical structure

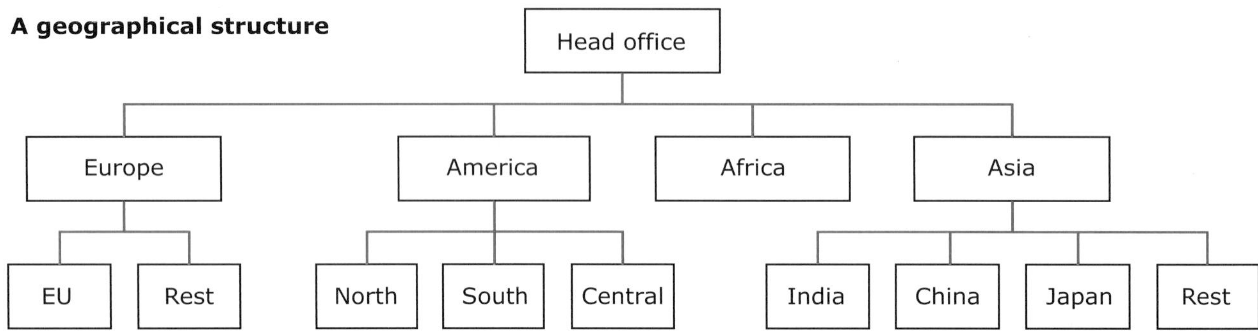

Figure 6 A geographical structure

A variation on the product or service structure is to group staff by physical location – for example, by region, country or continent. This has advantages for international organisations because there are likely to be big differences in markets, languages and cultures. National issues are usually best identified locally. However, there are several disadvantages to structuring by location:

- information flows between staff in different locations can be costly and problematic
- there may be duplication of activities, typically, support functions such as accounting, human resources and information technology
- it may be difficult to achieve integrated strategies across a number of different countries.

1.3.4 Matrix structure

We have seen that functional, product or service, and geographical structures all have disadvantages. In a matrix structure, each person has two reporting lines: (i) to the functional head; and (ii) to a project, product, service or region manager. These dual reporting lines are permanent. Advocates of matrix structures believe that they combine the advantages of both functional and product or service structures. Figure 7 is an example of a matrix structure.

In Figure 7, there is a Finance function and everyone in it will report to the Finance Director. However, each product (1, 2, 3, 4) will have its own independent finance team who also will report to the relevant product or service manager. Hence each group has two people to whom they report – a functional manager and a product/service manager.

There are, however, problems associated with a matrix structure. First, heads of reporting lines may need to meet regularly to decide how to apportion each person's time. What happens if the Finance Director and the Service 1 manager disagree about what the Service 1 finance team should be doing? Staff may feel uncomfortable with the change and uncertainty implicit in a matrix structure.

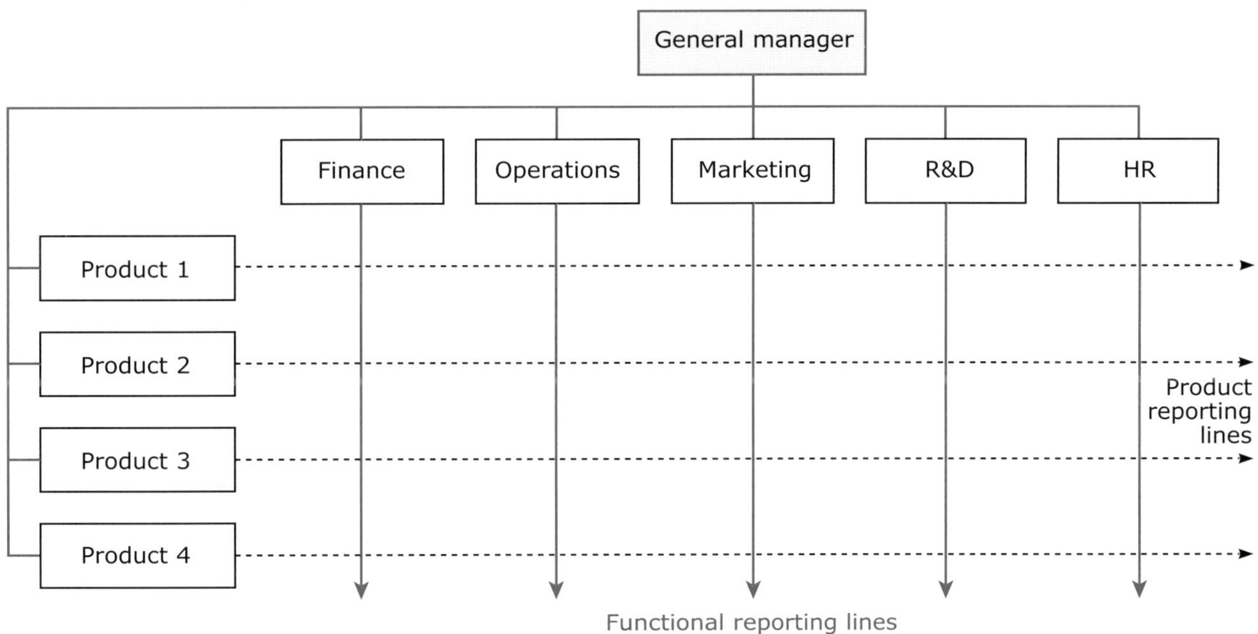

Figure 7 An example of how a matrix structure might work

In practice, matrix structures can be very difficult to manage. Dual reporting can lead to conflict, confusion, and overlapping responsibilities. This can then result in loss of accountability.

There could be a number of people who are responsible if Service 1 budgets are delivered late. The person responsible could be the Finance Director who changed the format, or the Service 1 manager who would not agree a sales budget, or the Service 1 finance team who took advantage of the conflict between the Finance Director and the Service 1 manager to get more time to do the task. Matrix structures can suffer from low responsiveness, slow decision making and high levels of internal political conflict.

On the other hand, they can offer a way of forcing people to work flexibly across functional boundaries (which can result in some productivity benefits for the organisation and outweigh the potential disadvantages). Organisations such as *Texas Instruments*, *Shell*, *NASA*, *NCR*, *ITT* and *Monsanto Chemical* have attributed some of their success to the matrix structure, which, in their opinion, helped them respond rapidly to customers' needs.

1.3.5 Project teams

Some organisations carry out the bulk of their work through project teams, which are often set up to react to changing circumstances and allow the organisation to respond quickly. They are an example of what are called **organic structures**, as distinct from the **mechanistic structures** that have been described so far.

Project teams draw staff from across the organisation, seconded on a full or part time basis. The latter can be very stressful for the individual, who effectively gets two jobs usually for the life of the project. An example of how this might work can be seen in Figure 8. This looks very like the matrix structure you have just looked at – and in a way it is. Clearly, those serving on project teams also have two reporting lines: to their functional head and to the project manager. The main difference is that the project teams are not permanent groupings like the Service 1, 2, 3, 4 groupings we saw in the matrix structure in Figure 7. Instead, these project teams only last as long as the projects on which they are working. This can lead to problems of control but, on the other hand, it can generate a tremendous sense of excitement as project goals are achieved. Also, fast moving changes can be implemented as a result of the flexibility and speed of response that this sort of structure can give an organisation.

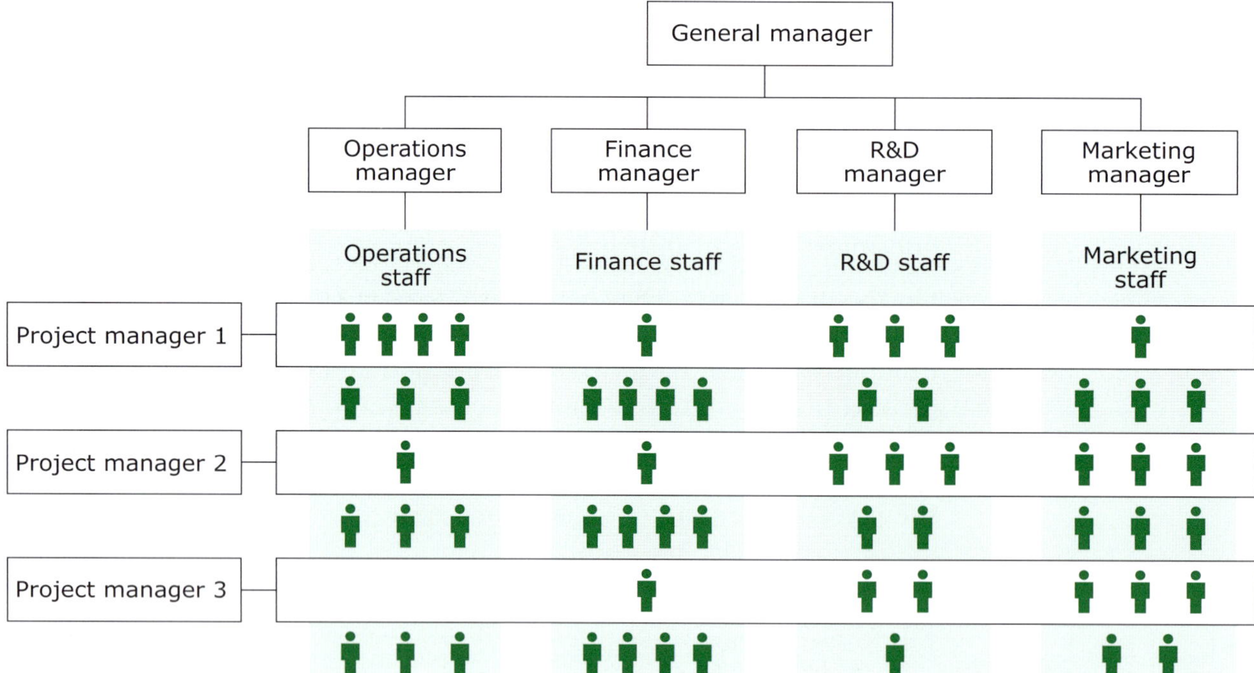

Figure 8 An example of project teams working across a functionally structured organisation

1.3.6 Hybrid structure

In practice, organisations evolve and changes to their structure should occur as and when necessary. This evolutionary process can produce structures that are well adapted to meeting the particular needs of the moment. A hybrid (mixture) of functional and product or service structures is common, and Figure 9 illustrates how this might look. Here we see that the sales function is structured along product lines.

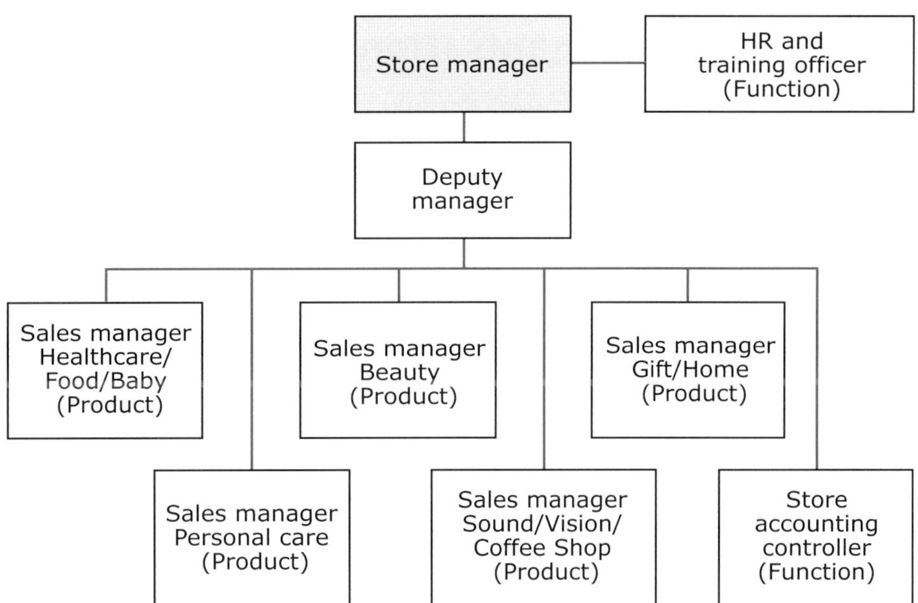

Figure 9 The organisational structure of a department store as a typical hybrid of product and function

A hybrid structure might include functional departments, the work of which is primarily internal (e.g., finance and HR), with the rest of the company organised on a product, service, geographic or project basis, as required to meet best the needs of the customer (as in Figure 9).

1.3.7 Recent trends in organisational design

In response to a dynamic, fast changing and competitive environment, flexibility and speed of decision making have become increasingly important in modern organisations. This has led to a number of trends in recent years:

1. The flattening of structures to remove levels in the organisational hierarchy. This shortens the chain of command and thereby increases the speed of decision making.
2. The establishment of multi-functional project teams and an empowered, multi-skilled workforce to increase flexibility.
3. A customer service orientation rather than an inward, internal process orientation.
4. The emergence of the 'flexible firm'. In an increasingly competitive environment, continuous reduction of costs is essential. Organisations have responded by replacing full time, permanent, salaried staff with temporary and part time contract labour. This allows greater flexibility, as labour can be readily taken on/laid off as demand conditions fluctuate. It also provides substantial savings in pension, health insurance and holiday pay costs. Such organisations consist of a small core of permanent full time salaried professional staff, who organise and direct the organisation's affairs, supplemented by various levels of contract staff. Some of these will be professionals contracted for particular projects, but the majority are likely to be the flexible workforce that carries out the routine operations of the organisation.

1.4 Organisational departments and functions

Management accountants are increasingly acting as 'business partners' to teams in other functional areas of the organisation. If they are to be accepted within these decision making teams, it is essential they understand the operations and technology of the organisation.

Business organisations typically consist of a number of departments or functions and it is important to have an appreciation of the purpose and activities of these departments/functions in order to understand the role of management accounting in the organisation. After all, management accounting is concerned with providing managers at all levels with information to help them undertake their various activities and to monitor/report on the impact of these activities on the organisation. A well designed management accounting system must be based on an understanding of what various managers actually do.

1.4.1 Typical business organisation departments and functions

A typical business organisation may consist of the following main departments or functions:

- Production
- Research and Development (often abbreviated to R&D)
- Purchasing
- Marketing (including the selling function)
- Human Resource Management
- Accounting and Finance.

1.4.2 The Production function

The Production function undertakes the activities necessary to provide the organisation's products or services. Its main responsibilities are:

- production planning and scheduling
- control and supervision of the production workforce
- managing product quality (including process control and monitoring)
- maintenance of plant and equipment
- control of inventory
- deciding the best production methods and factory layout.

Close collaboration will usually be necessary between Production and various other functions within the organisation, for example:

- Research and Development, concerning the implications of product design for production methods and cost
- Marketing, concerning desired product functionality, appearance, quality, durability and so on
- Finance, concerning the availability of funds for purchase of new equipment and the acceptability of inventory levels
- Human Resource Management, concerning staff motivation implications of job design and production methods.

Holding stock involves significant costs in the form of physical storage and warehousing costs, security, possible obsolescence and bank interest on the funds used to acquire or produce it.

Service organisations

Although many of the principles of good management in a manufacturing environment also apply in organisations that provide services (rather than manufacture products), service businesses, such as banking and professional firms of accountants and solicitors, do have a number of distinctive features which have implications for how they are managed.

1 Services are less easily standardised than manufactured products and so service quality tends to be more variable. This makes human resource management and motivation more critical.

2 Services are often 'intangible' (i.e., something that cannot be precisely measured or assessed) and multi-dimensional – what exactly is the 'service' being offered by a bank, a private hospital or educational establishment? This can make attracting customers more difficult as it often depends on promoting an intangible item.

3 Unlike manufactured products, services cannot be stored, but must be consumed as they are produced or they are wasted. This creates additional problems matching productive capacity with customer demand. This is reflected in, for example, the common practice of commercial airlines offering very cheap flights based on marginal cost to fill empty seats – a plane flying empty to New York is a service provided but wasted!

4 Ascertaining the cost of individual services is often also problematic, as the cost structure of many service businesses is such that costs are often shared among different services. This makes, among other things, pricing and the analysis of profitability of different services more difficult than with most manufactured goods.

1.4.3 The Research and Development function

The Research and Development (R&D) function is concerned with developing new products or processes and improving existing products/processes. R&D activities must be closely coordinated with the organisation's marketing activities to ensure that the organisation is providing exactly what its customers want in the most efficient, effective and economical way.

1.4.4 The Purchasing function

The Purchasing function is concerned with acquiring goods and services for use by the organisation. These will include, for example, raw materials and components for manufacturing and also production equipment. The responsibilities of this function usually extend to buying goods and services for the entire organisation (not just the Production function), including, for example, office equipment, furniture, computer equipment and stationery. In buying goods and services, purchasing managers must take into account a number of factors – collectively referred to as 'the **Purchasing Mix**', namely, Quantity, Quality, Price and Delivery.

- *Quantity*. Buying in large quantities can attract price discounts and prevent inventory running out. On the other hand, there are substantial costs involved in carrying a high level of inventory, as you will see in Unit 3 of this module.
- *Quality*. There will usually be a trade-off between price and quality in acquiring goods and services. Consequently, Production, R&D and Marketing Functions will need to be consulted to

determine an acceptable level of quality which will depend on how important quality is as an attribute of the final product or service of the organisation.

- *Price.* Other things being equal, the purchasing manager will look for the best price deal when procuring goods and services, although price must be considered in conjunction with quality and supplier reliability, in order to achieve best value, rather than lowest price only.
- *Delivery.* The time between placing an order and receiving the goods or services, the **lead time**, can be critical for production planning and scheduling and also has implications for inventory control. Suppliers must therefore be evaluated in terms of their reliability and capability for on time delivery.

In short, the 'purchasing mix' can be considered as making sure that the organisation has the right amount, of the right quality, at the right price, in the right place at the right time!

1.4.5 The Marketing function

In order to be successful, a business enterprise must either have a lower price than its competitors, or a product that is in some way superior – or both! A competitive strategy based on low price is known as a **cost leadership** strategy. A competitive strategy based on developing a superior product is known as a **differentiation strategy**.

Marketing is concerned with identifying and satisfying customers needs at the right price. Marketing involves researching what customers want and analysing how the organisation can satisfy these wants. Marketing activities range from the 'strategic', concerned with the choice of product markets (and how to compete in them, for example, on price or product differentiation) to the operational, arranging sales promotions (e.g., offering a 25 per cent discount), producing literature such as product catalogues and brochures, placing advertisements in the appropriate media and so on. A fundamental activity in marketing is managing the **Marketing Mix** consisting of the '4Ps': Product, Price, Promotion and Place.

- *Product.* Having the right product in terms of benefits that customers value.
- *Price.* Setting the right price which is consistent with potential customers' perception of the value offered by the product.
- *Promotion.* Promoting the product in a way which creates maximum customer awareness and persuades potential customers to make the decision to purchase the product.
- *Place.* Making the product available in the right place at the right time – including choosing appropriate distribution channels.

The historical evolution of marketing

Several writers (e.g., Harrison, 1978) have argued (and it is now widely accepted among management theorists and practitioners) that there have been three distinct eras in the history of advanced capitalist countries, such as the UK, which have affected the status, role and responsibilities of the Marketing function. These were:

1 **The Production Era** (pre–1930). This refers to a period of time during which products (and services) were relatively scarce (thereby constraining consumer choice) and the most important function of business was that of production. Marketing, in so far as it existed, was considered the least important function.

2 **The Sales Era** (1930–50). This refers to an era characterised by a shift in emphasis of business management from the production function to that of selling. With continued industrial development and innovations,

many new consumer oriented products became available and a much more competitive selling environment resulted. This made it necessary to seek out customers and make significant use of advertising, promotion and personal selling.

3 **The Marketing Era** (1950–present). This period marked another significant change in the attitude of senior management towards the status and responsibilities of marketing. This change, referred to by many writers as the **Marketing Concept** (Kotler, 1967), meant a departure from the previous concept of marketing as being the sales function of a business, to one where marketing had a much greater responsibility in total company policy formation and operation. Under the Marketing Concept, marketing was placed at the beginning of the process of determining the products (services) which were needed by the market, the price at which they should be sold and the way in which they were to be distributed.

'We produce an inferior line of goods. That's why we're looking for a real first class marketing man.'

1.4.6 The Human Resources function

The Human Resources function is concerned with the following:

Organisations are dependent on their employees. Consequently, their recruitment and selection require careful management.

- *Recruitment and selection.* Ensuring that the right people are recruited to the right jobs.
- *Training and development.* Enabling employees to carry out their responsibilities effectively and make use of their potential.
- *Employee relations.* Including negotiations over pay and conditions.
- *Grievance procedures and disciplinary matters.* Dealing with complaints from employees or from the employer.
- *Health and Safety matters.* Making sure employees work in a healthy and safe environment.
- *Redundancy procedures.* Administering a proper system that is seen to be fair to all concerned when deciding on redundancies and agreeing redundancy payments.

In recent years, the Human Resources function has attained a more important status as there has developed an increasing need (especially in service organisations) to 'get the most' from employees, in terms of customer service, for the benefit of the organisation.

1.4.7 The Accounting and Finance function

The Accounting and Finance function is concerned with the following:

- *Financial record keeping* of transactions involving monetary inflows or outflows.
- Preparing *financial statements* (the **income statement**, **balance sheet** and **cash flow statement**) for reporting to external parties such as shareholders. The financial statements are also the starting point for calculating any tax due on business profits.
- *Payroll administration.* Paying wages and salaries and maintaining appropriate income tax and national insurance records.
- *Preparing management accounting information and analysis* to help managers to plan, control and make decisions.

The role of the Accounting and Finance function will be considered in greater depth in Session 2 of Unit 1, where you will look at the role of management accounting in the organisation and how it fits into the wider Finance function.

1.5 Decentralisation and centralisation

Centralisation describes a situation where decision making authority is held predominantly by senior managers within an organisation. Such a situation is common within smaller businesses where the owner/manager takes all the important decisions. Centralisation, however, is not only found in such organisations, as a number of very large organisations, such as banks and some large retailers, are also highly centralised. Little, if any, discretion is given to branch managers, who must simply run their branches in accordance with the procedures established by Head Office.

Effective delegation has benefits for managers, staff and the organisation as a whole.

Decentralisation describes the situation where the authority to make decisions is delegated to people at lower levels of the organisation. This often occurs where growth in size and increased complexity make the delegation of significant decision making authority necessary. Decentralisation is a matter of degree and is usually present to varying degrees in most organisations.

Coates et al. (1996, pp. 115–6) identify the advantages and disadvantages of decentralisation as follows.

1.5.1 Advantages of decentralisation

- *Specialisation.* Managers can develop more detailed and specialised knowledge by concentrating on a limited aspect of the organisation's operations. This should result in better quality decisions.
- *Timeliness.* Quicker decisions are possible if it is not necessary to pass decisions up through the hierarchical chain of command. In addition, senior managers' time is then available for more important decisions affecting the future of the whole organisation.
- *Motivation.* Having authority to make decisions usually results in greater motivation and commitment and hence improved performance.

- *Human resource development.* Less experienced managers can 'learn their trade' without their mistakes jeopardising the entire organisation. The impact of mistakes/misjudgements is likely to be confined to a limited aspect of the organisation's operations.
- *Organisational segment performance comparison.* By dividing the organisation into separate segments, it is possible to evaluate which aspects of operations are performing well and which are not – something not usually possible when the inputs/outputs are at a more aggregate level.

1.5.2 Disadvantages of decentralisation

Managers' personal goals may differ from those of the organisation.

- *Dysfunctional decision making.* This occurs where managers take actions which improve the measured performance of their organisational segment, but damage the organisation as a whole. For example, a manager in one department may keep costs down in his/her own department in ways which have an impact on the quality of service provided to other departments.
- *Loss of control.* There is a danger that senior management may lose control of the organisation, as they become far removed from the detail of underlying operations and unaware of the decisions being made by lower level managers.
- *Increased cost of control.* Costly management information systems may be necessary to monitor the performance of lower management levels to ensure that delegated decision making authority is being used in the best interests of the organisation.

'In the interest of overcoming my reluctance to delegate, starting Monday I want you to do all of my worrying for me.'

1.5.3 Responsibility centres

Decentralisation results in the creation of separate **responsibility centres** – aspects/areas of the organisation's operations for which a particular manager is responsible. The main types of responsibility centre in common use are as follows.

Cost centre. The manager is assumed to be able to influence significantly the level of cost incurred and will be judged according to how well costs are controlled. There may or may not be revenues associated with the particular aspect of operations concerned, but if there are, the manager is assumed to have no control over these. Managers of cost centres will need regular information to be provided by the accounting system, concerning how individual cost items compare with budget – that is, **budget variance reports** (budgeting and budget reports will be considered in detail in Unit 4 of this module) and, of course, budgetary planning information, for example, cost targets to be achieved.

Revenue centre. The manager is assumed to be able to influence significantly the level of revenue earned and will be judged on the basis of this. Revenue centres are principally intended to be applied to sales operations, where the manager's responsibilities relate to the generation of income, whether or not there are attributable costs. Managers will need information concerning targets for individual revenue generating units (e.g., products), and also regular feedback information on actual versus budgeted revenues.

Profit centre. The manager is assumed to be able to influence significantly both costs and revenues and is judged on the basis of the level of profit generated by the particular aspect of operations concerned. The term profit centre is usually limited to the situation where the manager does not have responsibility for the level of investment in the centre (this decision being made by more senior management). Managers need information concerning both revenues and costs, for example, which products/services are profitable.

Investment centre. This is a type of profit centre, but one for which the manager also has significant influence over investment decisions. In such cases, it would be expected that, in addition to the normal profit centre measures (e.g., **Return on Sales**), profit would be related to the capital invested (e.g., **Return on Investment**). Managers need the same information as for profit centres and in addition, detailed appraisals of potential investments and control information concerning the level of investment (e.g., **working capital** levels, namely, inventory, receivables and payables) at any particular point in time.

Activity 1.4

- What do you think would be the appropriate type of responsibility centre for each of the following functions/departments?
  - A marketing department
  - A research and development department
  - A machinery service and repair department of a factory
  - The German manufacturing and distribution division of a large multinational company
  - A regional sales office of a US company
- How would you measure the performance of the manager/s of each centre?

Feedback

In some organisations, which are departmentalised on the basis of product groups, each product group department will have its own marketing activities. Where, however, an organisation is departmentalised on a functional basis and has a separate marketing department serving all its product groups, this department will incur expenditure on behalf of the whole organisation, but will not have its own revenues. It is likely then to be a cost centre.

The same is true of a research and development and a service and repairs department. Research and development is difficult to measure, but this is attempted in many firms by use of such measures as number of patents registered, percentage of sales from recently introduced products and so on.

The German manufacturing division of a large multinational company is likely to earn revenues (the level of which it can presumably influence) and incur costs (the level of which it can also influence). It is likely to be a profit centre, or even an investment centre.

A regional sales office is likely to be able to influence the level of revenue generated, but the main elements of cost (e.g., sales personnel salaries) are likely to be the result of decisions made by more senior (Head Office) managers. It is likely to be a revenue centre.

Responsibility centres should be evaluated according to financial performance (comparison of budgeted and actual costs, revenues and profits as appropriate). It will probably be desirable also to measure various non-financial indicators that are important for each centre achieving its goals – for example, in the case of the service and repairs department, average response time for machine breakdowns, and in the case of the research and development department, the number of new product or process innovations, and so on.

1.5.4 The divisionalised organisation

The ultimate form of decentralisation occurs when an organisation is separated into a number of investment centres that operate almost as independent businesses, each with its own profit responsibility. The basis of **divisionalisation** may be according to product (or service), geographical area and so on. Such a divisional structure is illustrated by the common legal form of groups with subsidiary companies. Each division in the organisation will typically have its own functional structure.

1.5.5 Line and staff relationships

The existence of an organisational structure implies that authority and control are exercised from above and pass down through the hierarchy. The relationships that result are known as **line relationships**. In any organisation, there should be a clear line of authority and responsibility from the top to the bottom of the hierarchy: the 'scalar chain' which indicates the line relationships. By contrast, **staff relationships** exist when a manager gives/receives advice from another organisational member. For example, the Accounting and Finance manager will provide information and analysis to help the Marketing and Production managers make decisions and control their respective operations.

Activity 1.5

This activity draws together some important ideas you have considered concerning organisation structure.

- Draw an organisation chart of the organisation in which you work, or one with which you are familiar. (Focus on the part of the organisation with which you are most familiar.)
  - What is the span of control?
  - How many layers or levels of management are there? Is your organisation tall or flat?

 - Is authority centralised or decentralised?
 - How clear is the definition of jobs? Do some of them overlap?

- To how many people do employees report? Are there single or dual lines of control?

You may find it difficult to draw the organisation chart, especially if there are dual responsibilities or reporting lines in some areas. You may find it useful to use dotted lines to represent working relationships that do not involve line management responsibility.

1.6 The organisational environment

Organisations exist in an environment – everything that surrounds the organisation physically and socially. The constituents of the organisation's environment are likely to have an important impact on the management of the organisation. An organisation's management must systematically analyse its environment in formulating plans to achieve organisational objectives.

The major environmental factors impacting on an organisation can be grouped under four headings: **p**olitical/legal, **e**conomic, **s**ocial/demographic and **t**echnological (reflected in the acronym: **PEST analysis**).

1.6.1 Political and legal environment

Clearly, there are laws that must be complied with, emanating from a number of sources, of which the organisation must be aware. These cover areas such as:

- ways of doing business (e.g., included within contract law); professional negligence (included within the law of tort)
- protection of consumers (e.g., *Sale of Goods Act 1979*, *Consumer Credit Act 1974, 2006*)
- safe working environment for employees (health and safety legislation)
- confidentiality and use of information held concerning customers or employees (*Data Protection Act 1998*)
- duties of directors and financial reporting requirements (company law, in particular, the *Companies Act 2006*)
- minimum wage, equal opportunities and unfair dismissal rules (employment law)
- pollution, waste disposal (environmental legislation)
- tax liabilities (tax law).

In recent years, the European Union (EU) has become increasingly important for member states in terms of international trade rules. In addition to requiring the removal of trade barriers, the EU requires that:

- there be free movement of capital between countries
- governments do not discriminate between companies in different EU countries in awarding government contracts
- financial services can be provided in any EU country

- telecommunications organisations be opened up to greater competition
- qualifications awarded in one country are recognised in the others.

In addition to the legal framework, government impacts directly on many organisations in a number of ways:

- via taxes or subsidies to discourage/encourage particular activities (e.g., alcohol consumption)
- national and, in particular, European, regulations have impacted on organisations in ways such as product standardisation, anti-discrimination legislation, workers' rights, etc.
- location incentives (often funded by the European Union) to encourage businesses to locate in particular areas
- by providing **barriers to entry** (e.g., the requirement to obtain a license to operate) and thus restricting competition in a particular field
- the government may be a major customer
- **anti-monopoly, competition legislation**
- as a supplier of infrastructure (e.g., roads), government can influence competition (e.g., road versus rail freight).

Political change, for example, wars, expropriation or nationalisation, political instability and so on, can also present a major threat to organisational plans.

1.6.2 Economic environment

Economic variables such as inflation, interest rates, savings patterns, economic growth, exchange rates, the levels of taxation and government spending all influence the amount of money people have to spend. This is likely to have an impact on most organisations. Businesses will experience, for example, varying levels of demand for their products or services and charities will experience varying levels of donations, as the amount of money people have to spend fluctuates in response to variations in major economic variables. The impact of economic variables on organisations is of such importance that a whole unit of this module, Unit 7, is devoted to this topic.

1.6.3 Demographic and social trends impacting on organisations

Demography is concerned with the study of data relating to the population and groups within it – births, deaths, diseases and so on – as indicators of the conditions of life in communities. Recent trends in the UK include a fall in the number of young people entering the labour market and an increase in the number of retired people.

Such data are used by organisations, especially in the identification of consumer markets, but they are also important for human resource management too.

Organisations must monitor demographic changes in their HR planning, especially with regard to recruitment. These changes must be taken into account in considering the best ways to tap into the future talent pool. It will be necessary to assess the current and future labour requirements in terms of both the level and the type of labour. The area from which the labour force is to be drawn must be identified. Once the labour force recruitment area is identified, the size and composition of the available labour force can be determined. The next step then involves a comparison of need as opposed to supply. In view of recent demographic changes, it is likely to be necessary to tap into currently under-utilised labour resources such as older workers and women returning to work after long periods spent bringing up children. HR policies need to be geared towards these requirements, for example, by providing training (e.g., in IT skills), part time working and child care facilities.

1.6.4 Sociological factors

Social class

A number of factors concerning the way society is structured will also be of relevance to many organisations. An important component of social structure is **social class**. Social class refers to the hierarchical distinction between individuals or groups in society. The factors that determine class vary from one society to another. The most obvious meaning of this term used to be in relation to the stratum of society into which one was born (aristocratic, middle class, working class) and class still means this to many people. In contemporary capitalist societies, it is common to define class in terms of income, status, education or profession, or a combination of these.

An important implication of the existence of such a class structure is that social class can create different customer groups. It is therefore a tool for **market segmentation**. The social classes are widely used to profile and predict different customer behaviour. Identifying a segment, in which customers share certain characteristics, such as level of income, is useful when developing products for those customers.

Culture

Culture refers to a set of shared attitudes, values, goals and practices that characterise a society or a social group within it. Cultural factors also have a significant impact on customer behaviour as culture is the most fundamental influence on a person's wants and behaviour. As with social class, culture can create customer groups and is therefore important for marketing. For example, there has been a cultural shift towards greater concern about health and fitness, which has created opportunities for serving customers who wish to buy:

- low calorie foods
- health club memberships
- exercise equipment
- activity- or health-related holidays.

Similarly, the increased desire for leisure time has resulted in increased demand for convenience products and services such as microwave ovens, ready meals and direct marketing service businesses such as telephone banking and insurance.

Each culture contains sub-cultures – particular groups of people with shared values. Sub-cultures can include nationalities, religions, and racial groups of people sharing the same geographical location. Sometimes a sub-culture will create a substantial and distinctive market segment of its own. For example, the 'youth culture' or 'club culture' has quite distinct values and buying characteristics from the much older 'grey generation'.

'This looks good. It's a six hour special on how society is becoming too sedentary.'

Cultural shifts also impact on other functions within an organisation. For example, there has been a dramatic increase in the number of women participating in the workforce and this raises a number of potential (and actual) discrimination issues, for instance, in terms of promotion and seniority and sexual harassment. HR policies must attempt to guard against these. Many organisations are responding by, for example:

- introducing flexible working hours to help women cope with the demands of career and family responsibilities
- providing education and training for managers and the workforce to encourage equal opportunities and discourage discrimination.

Another important cultural shift is the increasing concern for the physical environment among organisations' various **stakeholders** (employees, customers, investors, local community and so on). Organisations are responding in a number of ways:

- by introducing 'green products' to exploit the opportunity, for example, environmentally friendly deodorants, washing powder and cleaning agents
- by exercising greater care in disposing of industrial waste
- by reducing their **carbon footprint** by, for example, sourcing raw materials locally, not using air freight, using **video conferencing** instead of executive travel, using energy efficient appliances, etc.
- by **social and environmental reporting**. Increasingly, large companies are providing information (which is not currently required by law or accounting regulations) on environmental policies in their annual financial reports.

Many companies now present a corporate social responsibility report, although the content varies greatly between companies. This report is usually posted on the company's website.

In addition to such voluntary responses based on enlightened self-interest, organisations are increasingly needing to familiarise themselves with environmental legislation and direct government action, for example:

- the congestion charge in inner London
- fines for breaching pollution guidelines
- the landfill tax on hazardous waste.

1.6.5 Technology and its impact

In recent years, information systems (IS) and information technology (IT) have had a profound impact on most organisations. So pervasive is IT that IT skills are now essential for employees in virtually all organisations. IT has:

- facilitated flatter organisations and wider spans of control
- made possible faster, more accurate processing of larger volumes of data
- provided access to more – and more up to date – information for managers
- provided computer modelling (e.g., simulations) which can improve the quality of planning and decision making
- made possible the provision of control information to senior managers in real time
- improved customer service by provision of **electronic data interchange (EDI)** between organisations, customer **databases**, **extranets** and so on. These tools are discussed in more detail in Sessions 3 and 4 of Unit 1.

Perhaps even more dramatic than the improvements in information processing are the improvements in communications that IT has provided. **E-mail** provides instant worldwide messaging; **online conferencing** enables collaboration between people in geographically distant locations; **voice mail** allows communication between people whose working time schedules do not coincide; video conferencing allows face to face meetings without the need for expensive travel.

IT has also changed the relationship between employers and employees and the nature of work, for example in terms of:

- home working (**telecommuting**) is commonplace, providing more flexible working arrangements and reducing the time and cost of travelling to work
- greater visibility provided to management, allowing increased monitoring and control.

Activity 1.6

(a) Explain (spend no more than about ten minutes on this) why it is necessary for an organisation to undertake a PEST analysis.

(b) Give three examples of actions that might result from such an analysis.

Feedback

Environmental changes may have a significant impact on the organisation's ability to achieve its objectives, including, possibly, survival. Major environmental changes, however, can often be anticipated if the environment is systematically monitored and analysed, and appropriate action can be taken to safeguard against threats and/or exploit opportunities arising. Examples of possible actions in response to an environmental analysis include:

- developing a new product range based on emerging technology ahead of the competition
- developing staff education and training programmes in anticipation of future skills shortages in the local labour market
- planning a programme of redundancies based on anticipation of a serious economic downturn.

Summary

This session has provided an introduction to the nature of organisations. You should now have an appreciation of the way in which different organisations are structured and of the different components within an organisation. You should also have an understanding of the main environmental factors that impact on organisations, including political/legal, economic, social/demographic and technological factors.

Organisations are the context in which managers operate. In Session 2 you will go on to learn about the management functions of planning and control and the different levels of management. You will see why managers need to analyse the competitive environment in which an organisation operates and will consider some of the tools which can be used for this.

When you have considered the nature of management planning and control, you will be introduced to the nature and purpose of management accounting.

SESSION 2 Planning and control and the role of management accounting

Introduction

Upon completion of Session 2 you are expected to be able to:

- describe the management processes of planning and control
- explain the difference between strategic, tactical and operational management
- understand the different information needs at each management level
- explain the factors that influence the level of competitiveness in an industry or sector (using **Porter's five forces model**)
- describe the activities of an organisation that affect its competitiveness in terms of the **value chain**
- discuss globalisation and its implications for businesses and national economies
- understand the purpose and role of management accounting in an organisation.

In this session, you will look at the management functions of planning and control and at the different levels of management typically found in an organisation. You will also be introduced to some of the ways in which organisations analyse their competitive environment as part of the management process. All this is intended to give you an appreciation of what managers do, as a foundation for understanding the role of management accounting information and analysis – which is also introduced in this session.

2.1 What is management?

Management is often defined as consisting of planning, control and decision making.

Rational planning involves identifying objectives, activities and resources.

Planning is essentially a rational process. Planning consists of establishing objectives and developing plans to achieve those objectives. Control consists of making sure that the organisation stays on course to achieve its objectives, by taking appropriate corrective actions where planned and actual outcomes diverge (or modifying the plan if circumstances make this appropriate). Decision making consists of choosing between different possible courses of action based on a careful weighing up of the relative costs and benefits of each.

'Apart from appearing important, what other skills have you got?'

Activity 2.1

Your local community has decided to build a new hall in which to hold meetings and community events. A local landowner has offered to provide a site for the purpose and the Planning Authority has agreed that a hall can be built on the site, provided that the design is acceptable to them. A committee has been formed to oversee the project and has asked you to come up with a plan for taking it forward. The community is a small one and initial research has shown that the hall would need to cater for dances, meetings, adult education classes, indoor sports such as badminton, play groups for pre-school children and events that would be attractive to all sectors of the community. There are concerns about the additional cost of maintaining a new community hall and how the funds could be found.

You want to give the committee some idea of what the project might entail and decide to go through the first stages of a planning process so that you can be clear in your first report to them about what would need to be done.

Based on this limited information, spend about 15 minutes making notes outlining the objectives, the activities that would need to be undertaken, and the resources that might be available.

Feedback

At this early stage in the planning process, when all you are working from is an initial idea, you cannot be too specific about the activities that will be required or the resources. However, it does help to try to clarify the objectives. Your answers may have included the following.

Objectives:

- to build a hall that is versatile enough to meet the needs of all potential user groups now and in the future
- to build a hall that is attractive, energy efficient and easy to maintain
- to build a hall that meets the requirements of the local planners.

Activities:

- discover which appropriate skills are available in the community
- obtain an initial design for consultation purposes
- meet with the local planners
- research possible sources of funding for the project
- draw up plans for fund-raising.

Resources:

- professional and other skills that exist in the community
- grants from local companies
- grants from charitable trusts
- pledges of money from members of the community
- donations of materials for building purposes from local companies.

Anthony (1965), however, rejects this classification scheme, arguing that both planning and control involve decision making, which is not, therefore, a separate category of managerial activity. Instead, he describes management as consisting of planning and control at a number of levels:

1 Strategic management
2 Tactical management
3 Operational management.

2.1.1 Strategic management

Strategic management is undertaken by senior managers and is concerned with the formulation and implementation of strategy. **Strategy** has been defined by Mintzberg (1978, p. 934) as: 'a pattern or stream of decisions about an organisation's possible future domains'. Chandler (1962, p. 13) defined strategy as: 'the determination of basic long-term goals and objectives of the enterprise and the adoption of courses of action and the allocation of resources necessary for carrying out these goals'. Strategy is concerned with the general direction of an organisation. For example, in a commercial organisation, strategy is concerned with what products/services to provide, to which markets/market segments and the basis on which to compete, for example, price or product differentiation.

It is common for organisations' senior managers to undertake a **Strengths and Weaknesses, Opportunities and Threats (SWOT) analysis** as part of the process of developing strategy. This consists of, first, an internal analysis, identifying what the company's strengths and weakness, are regarding its resources and capabilities, compared with those of competitors and the needs of potential markets to be served.

This is followed by an external analysis identifying the opportunities and threats in the environment in which the organisation will be operating. These may be concerned with customers, competitors, technology, or the general economic, social, political and legal environment. The purpose of this external analysis is to identify opportunities to be exploited and threats to be safeguarded against if the organisation is to survive and achieve its objectives.

The SWOT analysis is a useful tool for helping senior management to select an appropriate strategy, for example, enter a new **product market**; reduce the scope of operations and specialise in **core business** areas; sell off unprofitable parts of the business; do nothing, etc.

Strategic management, then, is concerned with setting overall objectives, identifying possible strategies capable of achieving those objectives, evaluating these possibilities and selecting the strategy considered best able to achieve the organisation's objectives (given the strengths and weaknesses identified in the SWOT analysis). The emphasis is very much on the longer term and it is very much the province of senior management.

Levels of strategy

In the very common divisionalised form of company, consisting of a head office and a number of separate operating divisions, a distinction is made between 'corporate' and 'business unit' strategy. **Corporate strategy** is concerned with how the head office manages a portfolio of individual business divisions: which to invest in/divest from, which to close down and so on.

Business unit strategy is concerned with how an individual division competes in a particular product market area in order to achieve a long-term sustainable competitive advantage, either by being able to produce at a lower cost than, or by having a product offering that is superior to, the competition. The organisation's Marketing department will have a major contribution to make to this process.

In addition to corporate and business unit strategies, the main individual functions within an organisation will usually have their own **functional strategies**. For example:

- Production strategy will be concerned with sourcing of materials and components: whether to make or buy particular items; whether to apply methods such as **just-in-time (JIT)** and **total quality management (TQM)**; where to locate a factory and so on.
- Marketing strategy will be concerned with issues such as products/services to be offered, how they should be distributed, what the long-term approach to pricing should be, what the role of brands should be, etc.
- Financial strategy will be concerned with how to obtain and use the funds necessary for the organisation to undertake the activities required for achievement of its objectives, for example, whether to issue new shares or rely on long-term bank borrowing.
- Human Resources strategy will be concerned with how to attract and retain the right people, in the necessary numbers, in order to achieve the organisation's long-term goals.

JIT is a production system that aims to minimise and ideally eliminate stock, thereby reducing the substantial costs involved in holding large quantities of stock. Products are produced and materials bought as they are required rather than for stock. TQM is an organisation-wide management philosophy, embracing employee involvement and continuous improvement at all levels in the organisation. This contrasts with the traditional approach to quality control which was the responsibility only of the quality control department.

The competitive environment

An important task of management is the analysis of the competitive environment in which the organisation operates. Several analytical models have been developed by management theorists to assist in this process, one of which will now be described.

Porter's five forces model

Michael Porter (1985) has provided the 'five forces' model (illustrated in Figure 10) which can be used to help managers understand the nature and intensity of competition in a particular industry or product market sector. This indicates industry attractiveness and is thus a useful tool in formulating strategy (i.e., to help identify the business areas in which the organisation wishes to operate and the basis of competition). Industry profitability tends to be inversely correlated with the strength of competition.

According to Porter's five forces model, the level of competition in an industry is determined by five factors:

- the degree of current rivalry
- the relative bargaining power of suppliers
- the relative bargaining power of customers
- potential entry into the market by new competitors
- the availability of substitute products or services.

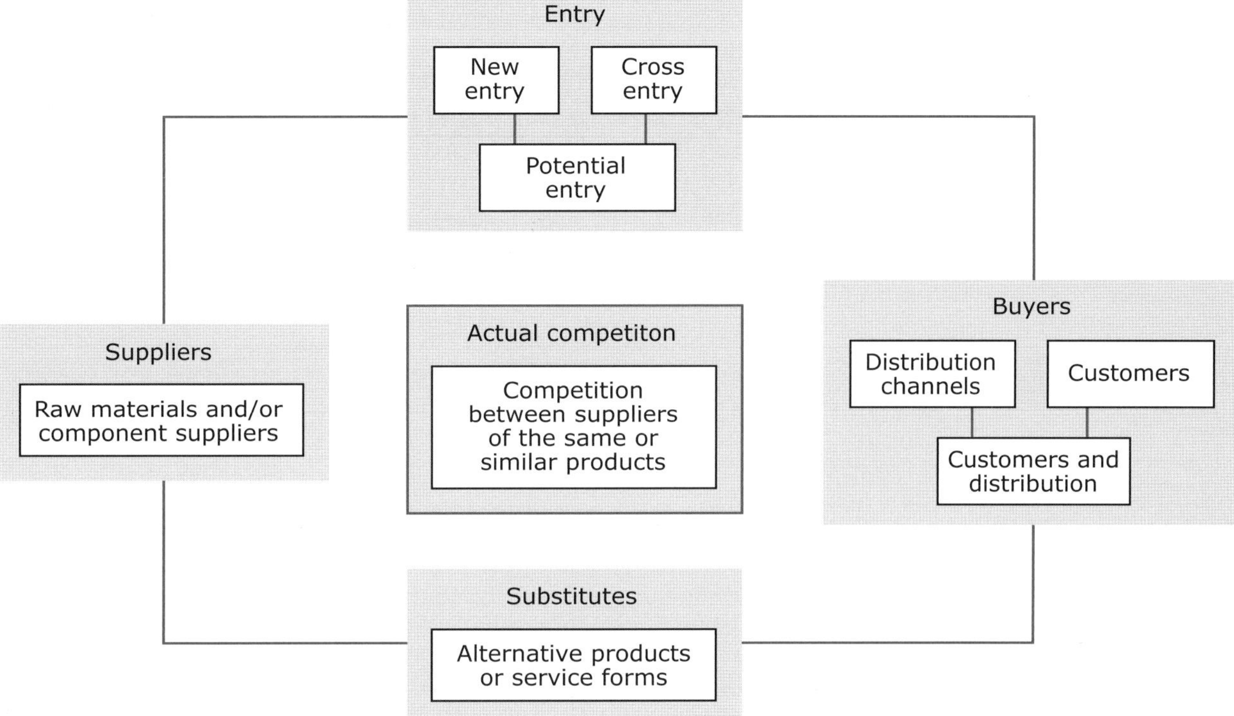

Figure 10 Porter's five forces model (1985, p. 5)

The degree of current rivalry can be assessed by looking at the market share of the main competitors. The more numerous the competitors, the more intense is likely to be the rivalry, especially price competition, with resultant low profit margins.

Suppliers are a weak or strong competitive force depending on their relative bargaining power. Supplier power may limit the negotiating power of firms in an industry if alternative sources of supply are not available. This will result in high prices and hence high costs/low profit margins for the purchasing firm. Alternatively, the power of suppliers is greatly reduced when the item supplied is a standard commodity, provided by numerous suppliers and for which the supply exceeds the demand.

The power of customers can also be a major competitive force. If either the distribution channel members or the final consumer of an item can exert major bargaining power (on price, quality, service levels, and so on), the profitability of industry participants will tend to be reduced.

An additional competitive force is the threat of entry into the market by competitors. Traditionally, this threat was conceived in terms of new firms. Increasingly, however, the major threat comes from established, well resourced firms crossing over from industries in which they are already established. This situation is referred to as 'cross entry'. To the extent that there is potential competition from new entrants to the market, there will tend to be downward pressure on prices charged to customers and profit margins will therefore tend to be reduced.

Finally, the threat of substitute products (as opposed to new entrants supplying similar products) places an upper limit on the prices a firm can charge and thereby on its profit margins. For example, clothes made from natural fibres face a serious threat from clothes made from synthetic substitutes.

Porter's model of industry attractiveness is relevant to both corporate and business strategies. In **portfolio management** (which is part of corporate strategy), certain competitive forces make one industry more profitable/attractive than another. Within a particular industry, such an analysis is helpful in selecting an appropriate competitive strategy (part of business strategy), for example, competing on price or on the basis of some sort of superior product offering.

Activity 2.2

For your own organisation, or one with which you are familiar, spend about 15 minutes identifying the competitive pressures of the industry or sector, in which the organisation is situated, using the five forces model.

Feedback

This activity has (hopefully) helped you to think about who your competitors actually are. It is not only commercial organisations that are faced with competition: charities must compete with other charities and with all sorts of other possible activities on which consumers may spend their money. Government departments must compete with other government departments for the limited funds available. Therefore, not-for-profit organisations must assess the competitive rivalries just as much as commercial organisations. The activity should also have helped you assess the relative bargaining power of those supplying your funds (e.g., customers) and those from whom you must obtain resources in order to operate. Together, these insights should help you form an overall assessment of just how attractive your industry or sector actually is.

The value chain

The **value chain** is another tool developed by Michael Porter (1985) to help organisations develop a competitive advantage through a better understanding of the activities through which they create value. Porter suggests that it is useful to separate a business into a series of value generating activities, which he refers to as the value chain. He argues that many businesses comprise a sequence of activities that can be classified as either primary or secondary activities, the latter existing to support the former.

The primary activities are:

- Inbound **logistics** activities, involving managing inbound items and including, for example, raw materials' handling and warehousing.
- Operations activities, involving the transformation of inbound items into products or services suitable for resale, for example, manufacturing and product design.

- Outbound logistics activities, involving carrying the product from the point of manufacturing to the buyer, for example, finished goods' warehousing and distribution.
- Marketing and sales activities, involving informing buyers about products and services and providing a reason to purchase. They include distribution and promotional activities such as advertising.
- Service, including all activities required to keep the product or service working effectively for the buyer, after it is sold and delivered. Examples include installation, repair, after sales service, warranty claims and answering customer enquiries.

The secondary activities are:

- procurement (purchasing)
- human resource management
- technology development (R&D)
- infrastructure (accounting and finance, strategic planning, etc.).

Any business's profitability depends on its ability to perform these activities efficiently and effectively so that the cost is less than the amount the customer is willing to pay for the outputs. A competitive advantage can be achieved by managing these activities better than competitors, either by having lower costs or by better differentiation (or even both).

A cost advantage can be achieved by gaining a better understanding of what drives cost and then reducing cost by managing the **cost drivers**. Differentiation may be achieved by focusing on the real value adding activities and performing them better than competitors. Examples of differentiation might include distribution systems that provide higher levels of customer service and purchasing of superior quality inputs. It is important to consider the linkages between value chain activities in order to create competitive advantage.

Alternatively the value chain can be reconfigured by, for example, **outsourcing** certain activities and concentrating on those activities where the organisation has particular expertise. Value chain analysis need not stop at the particular organisation's boundaries. Each business is part of a larger 'value system' as Porter calls it, consisting also of suppliers (and their suppliers) and customers (and, perhaps, their customers). Reconfiguring this value system by, for example, **vertical integration** or through strategic partnerships (e.g., JIT arrangements with suppliers) can also be an important source of competitive advantage.

Vertical integration can be contrasted with horizontal integration. The former consists of acquiring suppliers or customers in order to control the value chain. The latter consists of acquiring competitors in order to reduce competition.

Activity 2.3

Can you think of examples of how value chain analysis could help your organisation, or an organisation with which you are familiar?

Feedback

This activity should have helped you to understand better your organisation's strengths and weaknesses. It is ultimately the differences between your value chain and those of your competitors (think broadly about who your competitors are) that provide the source of any competitive advantage you may have. The activity has hopefully helped you think about which activities provide superior value for your ultimate customers and/or where you are able to control costs better, to gain a cost advantage by providing a given level of value at a lower cost.

Globalisation

A major development in the competitive environment has been 'globalisation'. Although globalisation is a process affecting technological, social, cultural and political phenomena, it is widely used to refer to economic phenomena. In this sense it refers to the integration of national economies into an international economy. This involves the removal of trade barriers between countries in order to allow the free flow of goods and services, capital and labour.

Globalisation has had a number of effects, including:

- the emergence of worldwide markets for goods and services
- the emergence of worldwide production facilities
- the emergence of worldwide financial markets
- the movement towards world government via institutions such as the **World Bank, International Monetary Fund (IMF)** and **World Trade Organisation (WTO)** which regulate the trade relationships among governments
- the increase in information flows between geographically dispersed locations
- a great increase in competition, requiring dramatic increases in productivity and innovation for survival
- the development of a global telecommunications infrastructure, using the Internet, communication satellites and so on.

As a result of these developments, managers must deal with a number of issues. In addition to the increased competition necessitating more rapid innovation and higher productivity/efficiency, globalisation has implications for human resource management. Workforces are increasingly diverse in terms of ethnic and racial background, gender and age. Managing diversity has become a major management issue. International production facilities, outsourcing, etc., have created a work environment where cultural diversity must be managed. Practices accepted in one culture may not be acceptable in another.

2.1.2 Tactical management

Tactical management is usually the province of middle managers and is concerned with converting the necessarily broad business unit strategies into more detailed action plans, allocating resources and responsibilities to particular areas in order to translate strategic plans into specific actions. Tactical management is thus:

> the mediating activity between strategic planning and the carrying out of specific activities. [It] involves monitoring activities and taking action to ensure that resources are being effectively and efficiently used in accomplishing organisational objectives.
>
> (Emmanuel et al., 1990, pp. 96–7)

Activity 2.4

The distinction between strategic and tactical management will usually apply to individual functions of an organisation. Consider, for example, the HR function, which may have strategic goals such as recruiting, developing and retaining high quality staff.

How might these strategic goals be reflected in tactical management plans?

Feedback

Tactical management might include planning salary surveys to ensure salaries offered are competitive, graduate selection and training programmes, employee attitude surveys and effective exit interviews (to ascertain people's reasons for leaving the organisation) and so on.

2.1.3 Operational management

Operational management is concerned with the implementation of tactical management plans. It is concerned with carrying out specific tasks on a day to day basis, for example, production scheduling, purchasing plans, placing advertisements in the media, and so on.

2.1.4 Types of information required at each level

At the strategic level, **senior managers** will require customised (i.e., non-routine), largely external and future-oriented information with a focus on the whole organisation, to help them with long-term planning. This is likely to include market information, competitor information and information concerning the general business environment (political, legal, economic, social and technological factors). Such information is likely to be in a summarised, broad and general form.

At the tactical level, **middle managers** will be concerned largely with information for monitoring and control. Such information will typically have a more historical and internal orientation than information prepared for senior management, consisting, for example, of budget variance reports. Tactical level information will be more short-term, (perhaps quarterly), internally oriented and be more detailed/disaggregated.

At the operational level, **front line managers** and supervisors will require information concerning specific tasks, purchase orders, production schedules, inventory control, and so on. Such information is likely to be almost entirely internal, very detailed/disaggregated and relate to the very short term.

2.2 What is management accounting?

Accounting information is central to the control function of management. This function is implemented by the continuous examination of documented evidence of organisations' numerical performance.

So far in this session you have learned about the activities with which managers are concerned. In order to undertake these activities and fulfil their responsibilities, managers require information, much of which will be provided by the organisation's management accounting function. The management accounting system (MAS) forms an important part of the organisation's overall **management information system (MIS)**, which will be considered in detail in Session 3 of this unit.

2.2.1 Common applications of management accounting

Some common applications of management accounting and the management processes they facilitate are shown in Table 1.

Table 1 Some applications of management accounting

Information/analysis provided by management accounting	Purpose/management function
Cost of making a product/providing a service	Establishing selling prices, product profitability analysis for product mix decisions (strategic management); cost control (tactical management)
Costs and benefits of different possible courses of action	Making short-term decisions about the most profitable use of existing resources (tactical management) and long-term capital expenditure decisions (strategic management)
Level of sales turnover necessary to cover full costs and prevent losses being made	**Profit planning (strategic management)**
Comparison of budgeted and actual costs and revenues	**Budgetary control (tactical management)**
Inventory levels, time taken to receive payment from debtors, time taken to pay creditors	**Working capital management** (i.e., managing the amount of short-term assets and liabilities such as inventory, trade debtors and trade creditors that the organisation has) (operational management)
Profitability of business segments (divisions)	Performance evaluation (strategic management)
Customer profitability	Identification and management of **key accounts (tactical management)**

The history of management accounting

Management accounting has its origins in cost accounting, which developed in the late nineteenth and early twentieth centuries. As industrial organisations increased in size and complexity, managers needed cost information for internal planning and control. Of particular concern was the ascertainment of product costs for pricing and cost control. The cost ascertainment methods that emerged were actually developed by practitioners in response to particular needs. A number of practitioners published their ideas and a widespread interest in 'cost finding' (as it was originally known) developed, culminating in the establishment of the Institute of Cost and Works Accountants (now the **Chartered Institute of Management Accountants**) in the UK in 1919 and the National Association of Cost Accountants (now the **Institute of Management Accountants**) in 1920 in the US. Cost accounting thereby became a separate field within the accounting profession, which previously had consisted of financial accounting and auditing.

Over the years that followed, new techniques and methods (e.g., budgeting, performance measurement, risk analysis, **marginal costing and contribution analysis** for decision making, and **discounted cash flow analysis**) were developed by practitioners, the professional bodies and eventually by university accounting departments. (All of these topics, as well as cost accounting, will be covered in later units of this module.) The wider body of knowledge that emerged became known as management accounting, as was reflected, for example, by the Institute of Cost and Works Accountants becoming the Institute of Cost and Management Accountants in 1972 and the Chartered Institute of Management Accountants in 1986.

Cost accounting is now a field, within the wider field of management accounting, that provides information about an organisation's costs, which may be used for both internal and external (financial accounting) purposes. When used for the latter, it is concerned with measuring the cost of sales (for the profit and loss account or income statement) and valuing inventory (for the balance sheet) in accordance with generally accepted accounting principles (GAAP) and/or specific accounting standards.

The prominence of management accounting varies between countries. The main techniques and methods seem to have been developed originally in the UK and US and have been adopted at different rates and to different extents in various other countries. In the English speaking world, management accounting tends to play a more prominent role in organisations than in many other countries, for example, Germany and Japan. This is also reflected in the professional status of management accountants across countries. In the UK, US, Canada and India for example, there are prominent professional associations of management accountants. In other countries such as Germany and Japan there are not such established institutions. Similarly, the position of 'management accountant' is widely found in UK/US organisations, whereas in Germany and Japan, such management accounting techniques and methods as are practised, tend to be practised by non-accounting personnel such as engineers or economists.

2.2.2 Management accounting and financial accounting compared

It is common practice to make a distinction between management accounting and financial accounting. The latter is principally concerned with preparing financial statements (income and expenditure accounts, balance sheets, cash flow statements) for external users such as those providing funds (shareholders, lenders), trade creditors, tax authorities and so on. These statements will also be of interest to managers of an organisation, but they will not be sufficient for managers' planning and control needs in the day to day running of an organisation. For this, much more detailed and more frequent accounting information is required. Providing such information and analysis is the function of management accounting. Table 2 compares the main features of financial and management accounting.

Table 2 Differences between financial and management accounting

	Financial accounting	Management accounting
Chief purpose	Production of summarised income statements and balance sheets by managers as a formal report on the stewardship of resources entrusted to them but should also, in the case of public companies, help interested parties (such as investors) make decisions. Depending on the type of the business entity, documents may be publicly available.	Production of detailed and up to date information used by managers to plan activities and control them. This information is not publicly available, but is internal to the entity producing it.
When information is prepared	Annually, at the end of an accounting period, but, depending on the type of business entity, may be every three or six months as well.	Normally prepared on a monthly basis.
Governed by	Legal requirements and often mandatory accounting regulations and/or conventions which may also dictate a required format (though this depends on the legal form of an organisation).	Management needs only, with no legal requirement to produce anything in any format, or anything at all! Information is produced in the format management deems most useful, e.g., by operating unit or product line, to record and monitor sales (by product, region, etc.), costs of production methods or products.
Perspective	Gives information about past performance, and might in practice be outdated by the time summarised documents are produced.	Comparative and up to date. While a given month's results are provided, these are usually accompanied by a total for all months to date, and comparative figures for a prior year, as well as for planned activities in the month and period to date.

Activity 2.5

A business is considering reducing the selling price of one of its products.

- What information will it require to decide whether this is the best course of action?
- If it does reduce the price, what information will the business need to evaluate whether it made the correct decision?

Feedback

In setting prices, the business should take account of the cost of producing the product (labour, materials and various 'overheads' such as rent and heating) and the estimated sales demand at a range of possible different selling prices. It will not usually be appropriate to set prices at a level that does not cover the full cost of production and earn an acceptable amount of profit. On the other hand, prices should not be set at a level beyond the value of the product as perceived by customers. Competitors' prices and the marketing department's view on likely changes in demand for given price changes will be important in assessing price-demand relationships.

After the price reduction, the business should monitor sales performance to see whether the anticipated increase in demand (presumably the reason for reducing price) was forthcoming. Also, were the actual costs of production as planned/estimated? Was the product profitability acceptable overall?

2.2.3 The role of management accounting in an organisation's management information system (MIS)

While it has been stated that management accounting is concerned with providing managers with information to enable them to carry out their planning and control (including decision making) activities, it would be wrong to suggest that management accounting provides all the information required by managers. The accounting system, comprising both financial and management accounting information is just part, although a very important part, of the organisation's total management information system (MIS). The MIS is the organisation's system, including hardware and software, for distributing information to managers to enable them to undertake their planning, control and decision making activities. Within the overall MIS, financial data will be collected and processed into useful information and provided to managers at various levels within the organisation. Financial and management accounting information will usually be prepared from the same basic data, but collated in different ways for the respective external and internal users. Both will draw in turn on the underlying cost accounting system.

It is now common for organisations to create a **database** (see Session 4 of this unit for a more detailed discussion of databases) of information which can be manipulated in various ways to serve the needs of different functions. Thus, for example, information on employees may be held just once on the database but extracted in different forms for human resource management and payroll processing purposes.

The MIS is likely to include:

- the accounting system
- production planning and scheduling information
- inventory records
- purchase and sales order processing records
- human resource management records, including employee records, staff appraisal details, etc.
- marketing data, including market research information, a customer database and so on.

2.2.4 Management accounting in commercial, charitable and public sector organisations

All organisations have limited resources with which to achieve their objectives. It is essential, therefore, that they use resources efficiently and effectively, and make the best decisions in the interests of the organisation. All organisations therefore require management accounting information and analysis. Thus, although management accounting has its origins in commercial organisations, it is now a part of all organisations of any significant size.

As indicated in Section 2.2.2, management accounting will be customised to the needs of managers in a particular organisation. Therefore, it would be expected that a traditional manufacturing organisation's management accounting system would be different from that of a service organisation. In many traditional manufacturing organisations, the core of the management accounting system is the cost accounting system, concerned with ascertaining unit product costs. The idea of a 'product unit' may not be a meaningful one in many service organisations, such as, for example, a bank or a university. In such organisations, the management accounting system emphasis is more likely to be on analysing costs by department or responsibility centre, for budgetary control purposes, rather than on product costing.

In the not-for-profit sector (charities and public sector organisations) the absence of the profit measure makes decision criteria (e.g., for capital expenditure/investment decisions) and performance measurement criteria (e.g., for evaluating organisational segments) are different from in the commercial sector. Nevertheless, decision making will still involve a careful weighing up of costs and benefits (however defined) of various different possible courses of action, and information concerning these (especially costs) is likely to be provided by the management accounting system. Part of the performance measurement of various organisational segments is likely to include financial criteria, again involving management accounting.

Budgets for planning and control are employed in virtually all organisations of a significant size, in all sectors. While profit centres may be absent from not-for-profit organisations, cost centres are commonly found and the principle of responsibility accounting applied just as in commercial organisations. The same type of management accounting techniques are widely applied in both commercial and not-for-profit organisations.

Summary

In this session, you have learned about the nature of management at different levels within an organisation and about the different information needs at each management level. You have also learned why it is important for managers to analyse the competitive environment in which the organisation operates and about some of the models that can be employed by them to do this. Finally, you learned about the nature and role of management accounting in helping managers perform their planning and control activities. In Session 3 you will be introduced to some basic principles of information systems which are important for the design and operation of management accounting systems.

SESSION 3 The role of information and information systems in organisations

Introduction

Upon completion of Session 3, you are expected to be able to:

- distinguish between data, information, knowledge and wisdom
- explain the attributes of good quality information
- understand how managers' information needs may be identified
- describe the main features of information systems used within an organisation
- identify the different sources of internal and external information useful to managers
- understand how information can be used as a competitive weapon.

As you have now seen, the management accounting system (MAS) is part of the organisation's management information system (MIS). The design and operation of an effective MAS therefore requires knowledge and understanding of the principles of information systems and of the opportunities offered by information technology. Such knowledge and understanding can enable the capture of relevant information from all relevant sources, to facilitate planning and control at strategic, tactical and operational levels.

3.1 Data, information, knowledge and wisdom

Knowledge management is the collection of processes that govern the creation, dissemination and utilisation of knowledge to fulfil organisational objectives.

Organisations are becoming more knowledge management-oriented, relying on knowledge as a key resource. It is important, therefore, for management accountants to be aware of the nature and role of information and knowledge and the principles of knowledge management, if they are to be an integral part of decision making teams in an organisation.

It is necessary, in understanding the nature of information systems, to recognise the difference between **data** and **information**, between information and **knowledge**, and between knowledge and **wisdom**.

Data

Data form a set of discrete, objective facts about events. For example, when a customer goes to a service station and fills the car's tank, that transaction can be partly described by data: when the purchase was made; how many litres were bought; and how much was paid. However, the data tell us nothing about why that service station was chosen and not another one, and cannot predict how likely that customer is to come back. Such facts say nothing about whether the service station is run well or badly, or whether it is failing or thriving.

Modern organisations usually store data in some sort of information systems. They may be entered into the system by a variety of data collection methods, described later in Section 3.1.1. At one time, the data were managed by central information systems departments that responded to requests for data and information from management of other parts of the organisation. Now data are less centralised and available on demand from desktop PCs, but the basic structure of what they are and how we store and use them remains the same.

All organisations need data and some industries are heavily dependent on them. Banks, insurance companies, utility companies and government agencies are obvious examples. Record keeping is at the heart of these 'data cultures' and effective data management is essential to their success. Efficiently keeping track of millions of transactions is their business.

Organisations sometimes pile up data because they are factual and therefore create an impression of scientific accuracy.

(Source: adapted from Davenport and Prusak, 1997, pp. 2–3)

Activity 3.1

What is wrong with the approach taken by the organisations described in the last paragraph in the previous box?

Feedback

The assumptions described are false on two counts:

(i) For many organisations and individuals more data are not always better than fewer.

(ii) Too much data can make it harder to identify and make sense of the data that matter. More fundamentally, there is no inherent meaning in data. Data describe only a part of what happened. They provide no judgement or interpretation and no sustainable basis of action. While the raw material of decision making may include data, data cannot tell you what to do. Data say nothing about their own importance or relevance. However, data are important to organisations – largely, of course, because they are essential raw material for the creation of information. When data are organised, patterned, grouped and categorised, they become information.

However, data are not homogeneous, but have many varied and different characteristics. Data are derived from invoices, credit notes, bank statements, purchase orders, income tax tables, surveyors' valuation certificates, cost calculations, economic statistics, hours worked, overtime records, production level records, inventory level records, customer order books, legal documents, etc.

The characteristics of individual items of data and the way they are gathered and communicated have a bearing on the process of making sense of the data, and on any undertaking involving measurement and data.

To be useful then, data must be transformed into information.

Information

Information is data endowed with relevance and purpose. This means that information is data that have been organised for some specific purpose that gives them meaning. In other words, when you take data and use them to produce something that is relevant to a specific purpose, what you have created is information.

Thus, unlike data, information has meaning. Not only does it potentially shape the receiver, it has a shape: it is organised to some purpose. Data become information when their creator adds meaning. We transform data into information by adding value in various ways. Let us consider several important methods, all beginning with the letter 'C':

- **C**ontextualisation: we know why the data were gathered
- **C**ategorisation: we know the units of analysis or key components of the data
- **C**alculation: the data may have been analysed mathematically or statistically

- **C**orrection: errors have been removed from the data
- **C**ondensation: the data may have been summarised in a more concise form.

(Source: adapted from Davenport and Prusak, 1997, pp. 3–4)

Information is, therefore, data presented in such a way as to be useful. However, information is only useful when someone learns from it, that is, when the information is absorbed and becomes knowledge. Let us move on and look at the next stage in the process, when information is used to generate knowledge.

Knowledge

Most people have an intuitive sense that knowledge is broader, deeper and richer than data or information. People speak of a 'knowledgeable individual', and mean someone with a thorough, informed and reliable grasp of a subject, someone both educated and intelligent. They are unlikely to talk about a 'knowledgeable' or even a 'knowledge-full' memo, handbook or database, even though these may be produced by knowledgeable people.

Knowledge is a fluid mix of experience, values, contextual information and expert insight that provides a framework for evaluating and incorporating new experiences and information. It originates and is applied in the minds of 'knowers'. In organisations, it often becomes embedded not only in documents or repositories but also in organisational routines, processes, practices and norms.

What this definition immediately makes clear is that knowledge is not neat or simple. It is a mixture of various elements. It is fluid as well as formally structured. It is intuitive and therefore hard to capture in words or understand completely in logical terms. Knowledge exists within people, and is part and parcel of human complexity and unpredictability.

Knowledge derives from information as information derives from data. If information is to become knowledge, humans must do virtually all the work. This transformation happens through such 'C' words as:

- **C**omparisons: how does information about this situation compare with other situations we have known?
- **C**onsequences: what implications does the information have for decisions and actions?
- **C**onnections: how does this bit of knowledge relate to others?
- **C**onversations: what do other people think about this information?

Clearly, these knowledge-creating activities take place within and between humans. While we find data in records or transactions, and information in messages (including reports), we obtain knowledge from individuals or groups of 'knowers', or sometimes in organisational routines. It is delivered through structured media such as books and documents, and person to person contacts ranging from conversations to apprenticeships.

(Source: adapted from Davenport and Prusak, 1997, pp. 5–6)

Now, let us look at the final stage in the process, namely wisdom.

Wisdom

Wisdom comes from having a correct understanding of the knowledge you have.

Let us consider an accounting example.

If you present qualified accountants with a company's balance sheet, they will be able to take that information and from it increase their knowledge about the company concerned. They will do so through knowledge creation activities, each of which starts with the letter 'C', like comparisons, consequences and connections. As a result of this process, they will know more than they knew before.

However, they have not been given the notes to the balance sheet, the auditor's report, or anything else other than the balance sheet. Their information is incomplete and, as a result, so is their knowledge. As long as they understand the new knowledge they have, and appreciate its limitations, they will have gained wisdom. However, if they behave as if they have an understanding of all aspects of the company that underlie a balance sheet, they may make a serious mistake.

Wisdom comes from understanding knowledge while, at the same time, being aware of the limitations of that understanding. In other words, true wisdom ensures that knowledge is used in appropriate ways.

'It's not whom you know or what you know that will help you succeed. It's what you know about whom you know, that you can hold over their head, that will help you succeed.'

Figure 11 presents these four concepts in a hierarchy that emphasises how each concept depends entirely upon appropriate application of the one that lies below it.

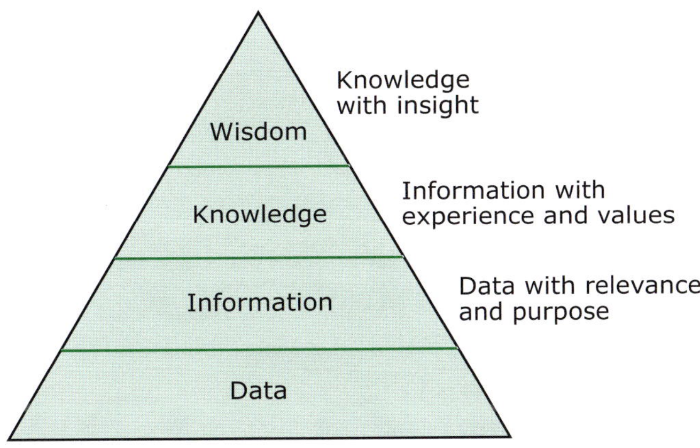

Figure 11 The data to wisdom hierarchy

Activity 3.2

Identify one example of each of the following that you may have:

- accounting data
- accounting information
- accounting knowledge
- accounting wisdom.

Feedback

Your list may have included some of the following.

- Data about: a sale, a purchase order, the inventory holding level of a particular item, a debtor. (You have the data but have not yet put them to use by converting them into information.)
- Information concerning: the payment record of a customer, the variable cost of a product, the net profit, the earnings per share of a company, the change in net assets over a period, the net book value of an asset, or, on a greater scale, any or all of the financial statements. (You have the information, but you have not yet converted it into knowledge as you have not yet absorbed the information.)
- Knowledge concerning: the credit rating of a customer, the cost of buying in a product that is currently being produced internally, whether one capital investment proposal is preferable to another. (You have absorbed the information and now know it: you have the knowledge.)
- Wisdom concerning: the circumstances in which to refuse a customer credit and the circumstances in which to grant the same customer credit, whether to overrule a recommendation that a capital appraisal project proceed, whether to invest in a business, etc.

3.1.1 Data collection methods

Before information is produced, data must be captured. There are various methods for capturing data for subsequent processing into information. The one chosen will depend on the nature of the task in hand, the relative efficiency and cost. Commonly used data collection methods using modern information technology include:

- magnetic ink character recognition
- optical mark reading
- scanners and optical character recognition
- document reading methods
- bar coding and electronic point of sale (EPOS)
- electronic funds transfer at the point of sale (EFTPOS)

- magnetic stripe cards
- smart cards
- touch screens
- voice recognition.

3.1.2 The characteristics of good quality information

There are four key characteristics which make information useful to users (Bentley, 1998):

1 *Relevance*. Information is relevant if it helps users make decisions. (Relevance also implies timeliness: information must be available when a decision needs to be made.)

2 *Reliability*. Information is reliable if it is free from material error and bias.

3 *Comparability*. Accounting information must be comparable with similar information produced by similar organisations and with information produced by the same organisation for different time periods. Comparability requires that similar items are treated in the same way for reporting purposes. For example, if one organisation **capitalises** part of its research and development expenditure (i.e., treats it as an **asset** on its balance sheet) and another writes it all off as an expense in the period in which the expenditure is made, the financial statements of the two organisations will not be directly comparable. Such comparability can be improved, however, if the **accounting policies** adopted are disclosed with the financial statements.

The International Accounting Standards Board (IASB) produces accounting standards which are intended to disseminate best practice, but also to facilitate comparison between the financial statements of different organisations by ensuring that similar items are treated in the same way.

4 *Understandability*. Information should be expressed as clearly as possible, and in a form that the anticipated users can understand.

Related to these is the concept of **materiality**. Information should only be included in reports to users of information if it is significant enough to affect economic decisions.

3.1.3 The information cost-benefit calculus

Providing information usually involves a cost as well as (hopefully) a benefit. Information should only be provided, therefore, if the benefits justify the costs, even if the other required characteristics of useful information are present.

It would not be surprising if you believed that organisations need an enormous amount of information. It has been suggested that data must be organised, patterned and grouped to become information and that to become knowledge, information must be contextualised, etc. Providing too much information is just as bad as not providing enough, because overload will prevent the development of potential knowledge. Like knowledge, information has to be managed. Information management is about providing managers with the right information.

3.2 Identifying the key information in an organisation

Technology allows organisations to collect a vast amount of information, but how can you ascertain the key information that managers need? No manager wants information which serves no useful purpose. It is, therefore, important that the information provided tells managers what they need to know. What is it that managers need to know? They need to know those things that are critical to the success of their role and, ultimately, critical to the success of the organisation. Such 'things' are known as **critical success factors (CSFs)**.

3.2.1 Critical success factors

CSFs can be thought of as being what you need to be good at in order to achieve your ultimate aim. Managers need to identify the CSFs relevant to their job and organisation and need to be provided with information that will enable them to assess whether or not they are achieving their CSFs. Research has shown that most managers consistently use only about five or six CSFs to monitor effectiveness, so a large volume of information is not normally what a manager needs.

The following activity concerns how you would go about identifying CSFs.

Activity 3.3

Imagine you are a manager in any organisation you know well, perhaps, but not necessarily, an organisation for which you have worked or are currently working.

How would you identify the CSFs that you need?

Feedback

To derive the CSFs for an organisation, it is useful to start from the objectives, such as what it wants to achieve, and then decide what it needs to be good at in order to achieve those objectives. This will then help determine the information (and **performance indicators**) needed in order to measure whether the managers and their organisations are going to achieve their objectives.

Rockart (1979) identified four different approaches.

1 The null approach. Managers have diverse, changing and unpredictable needs, and if you asked them what they wanted they invariably would not know. Not surprisingly, in those circumstances, the response from information systems analysts was to do nothing.

2 The by-product approach. Management information was produced as a by-product of the transaction processing system. This was often in the form of crude, unfiltered and unprocessed computer print-outs, frequently simply pages of transaction listings, being sent out as management information, supposedly to help decision making. These reports were often ignored as unintelligible or useless. Perhaps worse, managers spent many hours trying to identify for themselves relevant information among the mass of data they received. The low cost of producing these reports was more than compensated for by the high labour cost of managers working through them in search of useful information.

3 The total study approach. This is the opposite of the by-product approach. The IT department carried out a complete, but far too extensive, study of information needs that usually resulted in the conclusion that managers needed everything.

4 The key indicator approach. Under this approach, the information systems generate **exception reports** on specific **key performance indicators**. It prevents information overload and the information is specifically targeted to the needs of managers.

Activity 3.4

Which of these four approaches do you think was the one most likely to provide useful information to managers?

Feedback

The key indicator approach was found to be the most useful approach of the four. This leads to another question you should think about, but which we will not consider here, namely, why were the other three approaches used at all?

Activity 3.5

Think about an accounting firm. Suppose it has an objective of 'increasing market share over the next five years to become the market leader as measured by total revenue from audit fees'.

What do you think the key indicators (or CSFs) would be in achieving this? What needs to be happening if this objective is to be met?

Feedback

You may have come up with some of the following:

- maintaining client loyalty
- obtaining new clients
- extending the availability of the accounting firm's audit services to a wider geographical area
- providing a high quality service at competitive fee levels
- finding innovative ways of adding value to customer service
- maintaining a well trained, client-focused workforce.

Of course, you may have come up with other ideas. It does not matter, as long as you are following the general idea. In reality, you would find that members of an organisation will come up with different ideas and that, through a process of discussion, they will ultimately narrow the list down to the key indicators.

Exercises like this can be a valuable way of getting to grips with the subject. Apply the exercise to an industry in which you work or have worked, or to a bank or a hospital. Are the key indicators the same?

As you may have realised, you will find that the key indicators will change depending upon what you are trying to achieve.

Activity 3.6

Spend about ten minutes thinking about the information that would be needed to monitor whether the accounting firm in Activity 3.5 was, in fact, being good by reference to key indicators (CSFs) identified for success.

Feedback

Some information to be monitored might be:

- fees by service level, by office
- hours worked compared with hours charged to client
- average enquiries per week for other services

- average client retention period
- growth rate of client base
- the number of client complaints per 100 clients
- market share
- demographic analysis of population in areas where offices are situated
- percentage of local businesses that use the office
- competitor analysis
- external perceptions of the firm.

One of the things we hope you notice about this is that not all the information is internal to the organisation and, therefore, in this case the collection of external information is important and should not be ignored.

CSFs are derived from the concerns of senior management and cover such areas as industry trends, market positioning and the wider business environment.

3.2.2 Possible sources which can inform the determination of CSFs

The following sources can inform the determination of CSFs:

- the industry in which the business operates
- the firm itself and its position within the industry
- the organisation's wider economic and social environment
- organisational factors which are a current management concern, for example, high inventory levels or low morale among staff.

Even though organisations compete in the same market place, if their competitive strategies are different, the CSFs will be different, because the objectives will be derived from a different competitive position. For example, Rolls Royce and Toyota are both automotive manufacturers, but are unlikely to share many of the same CSFs because they compete for customers in different markets.

So, how do organisations hone down the amount of information with which managers are provided? One possible approach is to ask users to define their CSFs by:

- listing the organisation's, department's or individual's objectives
- determining which factors are critical for accomplishing the objectives
- determining a small number of primary information measures for each factor.

This approach will allow the information to be tailored to each manager's own requirements. However, you must remember that some managers may require similar information, but use it for a slightly different purpose. Therefore, the technology should allow the organisation to utilise the data it has available in order to provide a tailored view to each user, specific to their individual needs. In other words, IT becomes the enabler to providing the information.

'We've broken your list into eighty four sub-groups. Our work here is done.'

3.2.3 Information as a means of competitive strategy

It is possible to compete in the market by using information as a resource. Take the example of a florist who received a frantic telephone call from a man who had only just remembered his mother's birthday and requested some flowers to be sent immediately to her home. The florist did this without any problem at all. However, she kept the details on a computer and the next year, a few days before her birthday, the florist rang the man to see if he wished to send some more flowers. Pleased with the reminder, he responded positively. The florist developed a database of all the customers and, where possible, the reason for the flowers being purchased. In this way the business was able to use the information to build up future custom. This example of using information, rather than just the technology, again illustrates the importance of information to business.

Information and technology also come together in the loyalty schemes being developed by large retailers. Each transaction produces information about the purchasing patterns of particular customers. In return for the reward from the bonus scheme, the customer is providing a great deal of information to the retailer who can then process this data into detailed knowledge about the buying habits of individual customers. The sheer volume of the data available has delayed the deployment of these schemes, but it is now possible for companies to target customers with discounts and offers which reflect their personal shopping preferences, so that, for example, vegetarians are not offered discounts on meat products.

3.3 Information systems

Management information systems traditionally concentrated on reporting historical transaction-based data for control purposes only. Over the years, we have seen a variety of computer applications developed that help target information to the various levels and types of decision made within an organisation.

3.3.1 Characteristics of information

To appreciate the differences in computer applications that have developed, it is useful to review briefly the characteristics of information at the three main levels of decision making in the organisation. From Session 2 of this unit, you should recognise these as the operational level of the core activities, the tactical level of middle management and the strategic level of senior management.

Each level has different characteristics in terms of:

(a) time horizon
(b) decisions made
(c) level of detail
(d) scope
(e) format
(f) degree of uncertainty
(g) source of data.

Table 3 Characteristics of information at the three levels of decision making.

	Operational	Tactical	Strategic
(a) Time horizon	Short-term, historical	Medium-term; annual; 2–3 years	Long-term future 5–10 years
(b) Decisions made	Day to day control	Semi-structured, tactical	Unstructured strategic decisions
(c) Level of detail	Detailed, 100% accurate	Approximation	Summarised reports
(d) Scope	Department, group, individual	Divisional or departmental	Covering the whole organisation
(e) Format	Routine standard report	Routine, some ad hoc	Ad hoc, graphical data, some routine data incorporated
(f) Degree of uncertainty	Historical accuracy	Moderate degree of uncertainty	High degree of uncertainty
(g) Source of data	Internal	Internal, some external	External, but some internal

Each of the levels of decision making is supported by information systems that reflect the nature of the decision processes.

3.3.2 Operational systems (or transaction processing systems)

These are the main transaction-based processes that computers have traditionally been allocated to, for example, sales order processing and distribution. They are key to most organisations and have benefited from computerisation, owing to the increased speed and accuracy of processing that computers offer.

However, by nature, such systems have pre-defined functions. They represent high volume, low value transactions, and are concerned with efficiency. They are inflexible towards particular and specific information requirements presented on an ad hoc basis.

A transaction processing system may employ batch or online processing. In a **batch processing system**, transactions are stored together in **transaction files** and then processed together in batches at a later date. This mode of processing is common for applications such as payroll. It is appropriate where there is a large volume of transactions and it is not essential that the system is up to date all the time. For example, with payroll processing, the end of the month is adequate. In an **online processing system**, users enter information directly and the system is maintained with information constantly up to date. This mode is appropriate where it is essential that records are continuously up to date, for example, in a travel agent booking airline flights where the number of seats available must be continuously monitored.

3.3.3 Decision support systems

Decision support systems (DSS) can be used at any of the three levels. Examples at each of the three levels could be (operational level) inventory re-ordering; (tactical level) make or buy decisions; and (strategic level) capital investment appraisal. They may be capable of complex mathematical modelling to simulate the behaviour of an organisational component in an unpredictable situation. They are a support tool as people make the final decision, but they guide the decision makers towards a more informed decision than might otherwise be the case.

DSSs are able to produce results based on a model representing the relationships between the variables in the system. They are flexible and often include the facility to undertake **sensitivity analysis**. This means that 'what if?' questions can be asked by the decision makers, thus increasing their information and enabling them to establish the constraints that apply to the variables that are being analysed.

Spreadsheets running on a PC can be used to provide simple DSSs, but specialised software packages and custom-built systems are used for more complex simulations.

3.3.4 Executive information systems

These are computer-based systems that deliver information to strategic level managers. An **executive information system** (EIS) can be regarded as a special type of DSS. The key features of an EIS are that it:

- is organised around CSFs
- provides information at different levels of detail (i.e., users can 'drill down' from summarised information to ever-increasing levels of detail, sometimes as far as the original data)
- includes a sophisticated user-friendly graphical user interface to help its ease of use
- has a fast response time and is accessible from many locations
- is tailored to each individual manager's style
- makes extensive use of external data
- shows trends, ratios and deviations.

Managers at this level are primarily looking for a view of the organisation and are carrying out long-term planning. They are interested in trends. Many of the decisions taken have a high degree of uncertainty attached and much of the information they use is from external sources. Therefore, their need for information is different from those operating at the operational and tactical levels.

'I have a very important meeting with the shop floor this afternoon, Miss Wilkes. I need you to find out what it is we do.'

3.3.5 Office automation systems (OAS)

Office automation systems (OAS) are networked computer systems **local area networks (LANs)** that support office work for handling and managing documents and facilitating communication. Applications include word processing, e-mail, teleconferencing and intranets, spreadsheets, presentation packages (e.g., Powerpoint), personal databases and note taking systems (e.g., notepad) are also considered part of OAS.

3.3.6 Knowledge work systems (KWS)

Knowledge work systems (KWS) are computer applications that support knowledge workers by helping to create new knowledge. **Computer aided design (CAD)** and **Computer aided manufacturing (CAM)** are good examples of KWS. KWS are also widely used in the financial services sector to support trading and portfolio management.

3.4 Sources of the information

Most of the information, if not all of it at the lower levels of management, will come from internal sources. However, much of the information at the senior level is external. The phrase **environmental scanning** is often used to describe the process of gathering external information, which is available from a wide range of sources.

Activity 3.7

Spend five minutes writing down as many possible sources of external information as you can think of that an organisation could use and the types of data available from each one you think of.

Feedback

Those sources might include the following.

(a) Government. Local government bodies frequently provide information and assistance, particularly to encourage new business ventures.

(b) Advice or Information bureaux. Examples include Consumer Standards Offices, Offices of Fair Trading, Law Centres, Tourist Information Bureaux and so on. These provide enquirers with information in their own particular field, in the form of advice, information leaflets and fact sheets.

(c) Consultancies. General market research organisations, like MORI and Gallup, are able to provide organisations with information of a general nature as well as undertaking specific projects for them. Specialist market research companies provide market intelligence for specific industries.

(d) Newspaper and magazine publishers. The quality press is a vital source of general economic information. Most industries are also served by several trade journals and magazines.

(e) Specific reference works. There may be specific reference works which are used in a particular line of work. Legal firms, for example, refer constantly to law reports.

(f) Libraries and information services. These may be part of the free public library system, or associated with a learned professional institution, such as the Institute of Chartered Accountants in England and Wales, or the Law Society, or an academic institution.

(g) Increasingly businesses can use each other's systems as sources of information via electronic data interchange (EDI – see Section 3.4.1), which involves the exchange of routine business documents between the computers of suppliers and their customers.

(h) Electronic sources of information are becoming ever more important.

- For some time there have been TV-linked 'viewdata' services which offer a very large bank of information gathered from organisations such as the Office for National Statistics, newspapers and the British Library. Some companies, such as Reuters, operate primarily in the field of provision of information, whereas others, such as Topic, specialise in stock market information.
- Information can be found on virtually any subject via the Internet using a browser such as Internet Explorer and a **search engine**.

3.4.1 Electronic data interchange

Electronic data interchange (EDI) is a form of computer-to-computer data interchange. The general concept of having one computer communicate directly with another might seem straightforward enough in principle, but in practice there are major difficulties.

The EDI concept developed in industries where large volumes of paper transactions were being sent between organisations. Typically, car component firms were dealing with assembly plants and the orders and invoices were numerous and time-consuming to process. It became much more economical to send all this information electronically.

Organisations are becoming increasingly aware of the importance of the supply chain (e.g., the links between supplier, manufacturer and retailer) and its relationship to the notion of a value chain. In particular, as computer systems and software have become more complex, many larger organisations have direct links from their information system into the information systems of their suppliers and their customers, thus maximising the efficiency of both the purchasing and selling function.

With the rapid development of electronic commerce on the Internet, secure data transfer technologies from firm to customer have been developed. These allow smaller companies to offer economical, small-scale EDI, including online ordering.

Summary

This session has considered the importance of information in an organisation and the ways it may be organised (in the form of various information systems) to help managers plan, control and make decisions. It has also considered how information needs differ at different levels in the organisational hierarchy. Session 4 will consider how information technology can be exploited to enhance the flow of information through an organisation.

SESSION 4 Information management and information technology

Introduction

Upon completion of Session 4, you are expected to be able to:

- explain the concept of a database
- describe the function of a database management system (DBMS)
- describe the concept and use of data warehousing and data mining
- explain the concept of information markets and the growing use of the Internet and its impact upon businesses
- describe concerns over security and understand some of the remedies for these
- explain the use of intranets
- explain the meaning of 'knowledge work'
- identify whether technology will help in a particular situation
- describe the use of expert systems and the benefits organisations can derive from their use.

In the previous session, you looked at the role of information in organisations, the main sources of information and at the main features of information systems. Since management accountants are concerned with providing information, it is important for them to have a good understanding of the principles of information management and the opportunities offered by information technology. In this session you will learn how information and communications technology can be exploited in the collection, storage and distribution of information within an organisation. This session also looks at how information and knowledge can be used to enhance the value added by an organisation, but focuses

more on the aspect of managing large volumes of data and exploiting them via the use of technology. The technology in this instance should be viewed as the enabler. It is the mechanism that enables you to do things differently, or collect and analyse vast amounts of data.

4.1 Databases

A database is a store of data. Ideally, each data item should be held once and once only within a database.

Activity 4.1

Why do you think each item should be held once only?

Feedback

If you enter an item more than once you risk having different entries for the same item. For example, imagine your database contains a column for inventory and a column for purchases made in the year. You purchase 12 items of inventory and enter that number in the inventory column. You also enter '12' into the column for all purchases made in the year. You then return three items of inventory and enter this number in a column for returns made to suppliers in the year. However, you forget to enter 'minus three' in the inventory column. Your database now shows two different amounts relating to that purchase of inventory. To complicate things further, you may have entered '12' into the inventory column and '21' into the purchases made for the year column – one transaction, but two different records. In some well constructed modern database systems (see Section 4.1.2 on ERP systems), the entry of the purchase would automatically increase the inventory, as the various functions of the business are automatically linked via the computerised modelling system.

4.1.1 Database management systems

In its strict sense, a database is a collection, or file, of data structured in such a way that it may serve a number of software applications without its structure being dictated by any one of those applications. The concept is that programs (applications) are written around the database to use its data in various ways, rather than files being structured to meet the needs of specific application programs. For example, accounting transactions such as sales may be stored in the database and then used by sales and marketing personnel to analyse sales by geographical area and by accountants to provide a list of aged debtor balances (age analysis of receivables) for credit control purposes.

The term database is also rather loosely applied to simple file management software.

As shown in Figure 12, the **database management system (DBMS)** manages the interface between the data and the application programs, such that the data can be made available to other users, providing them with their own tailored view.

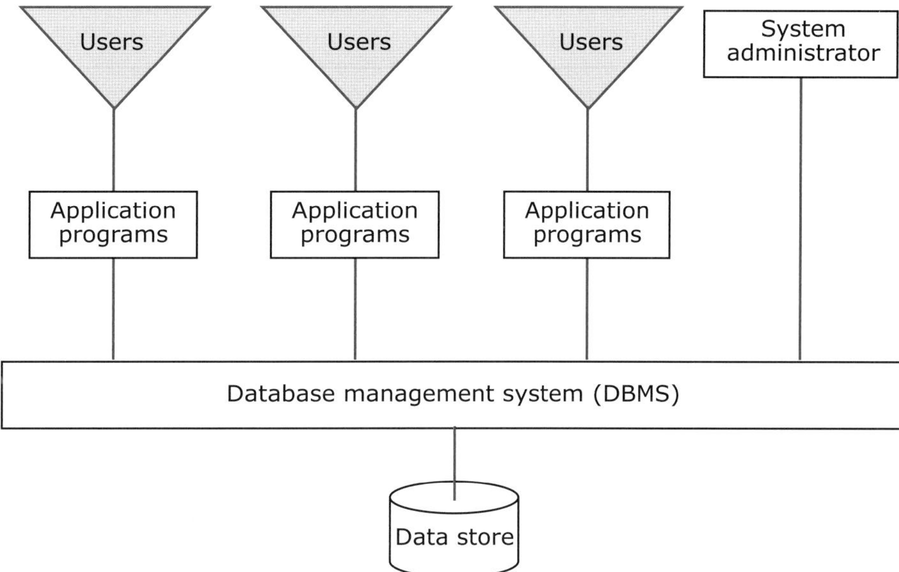

Figure 12 The database management system (DBMS)

An example of a database bringing benefits to an organisation is given by Otis Elevators.

Otis elevators

Customers had been complaining that when elevators (lifts) broke down, engineers often did not know about that particular model. This increased the repair time. Otis decided to establish a database to enable the company to track the service life of the lifts installed throughout the USA.

By using the database, Otis managed to isolate which type of lift typically developed certain types of fault. This enabled the company to train the engineers on specific types of lifts and use the information to redesign the lifts to prevent the faults occurring. By analysing the database geographically, the company built up a picture of the distribution of lifts by type. For example, lifts in the city centre tended to be larger, covering more floors than those in small town buildings and on the outskirts of the cities. Using this information the company improved its training programme for engineers by training them about the type of lifts which were common in the area where the engineers were based. This also allowed the company to obtain a better distribution of spare parts in the regional depots.

Eventually the engineers were directed to work from home and hold spares in a mobile toolbox or van. This meant that the regional depots were reduced in number with resultant cost savings.

However, the engineers began to complain of feeling isolated and missed seeing their colleagues at the depots. The company gave them mobile phones and allowed them to talk to each other, about anything: it did not have to have anything to do with their work! It also gave the company the ability to contact engineers when a customer called in their area and achieve a fast response time. This resulted in happy customers, happy engineers and thus happy shareholders.

In this example, it is the engineers themselves who create most of the data.

Activity 4.2

Using this example, spend five minutes writing down the wider aspects of the organisation that it demonstrates, such as staff issues, organisational issues and competitive position.

Feedback

Many of the wider aspects of the organisation that the example demonstrates can be seen in Figure 13, which shows how the introduction of the database, together with all the various changes to staff and the organisation, led to improved performance in the market place.

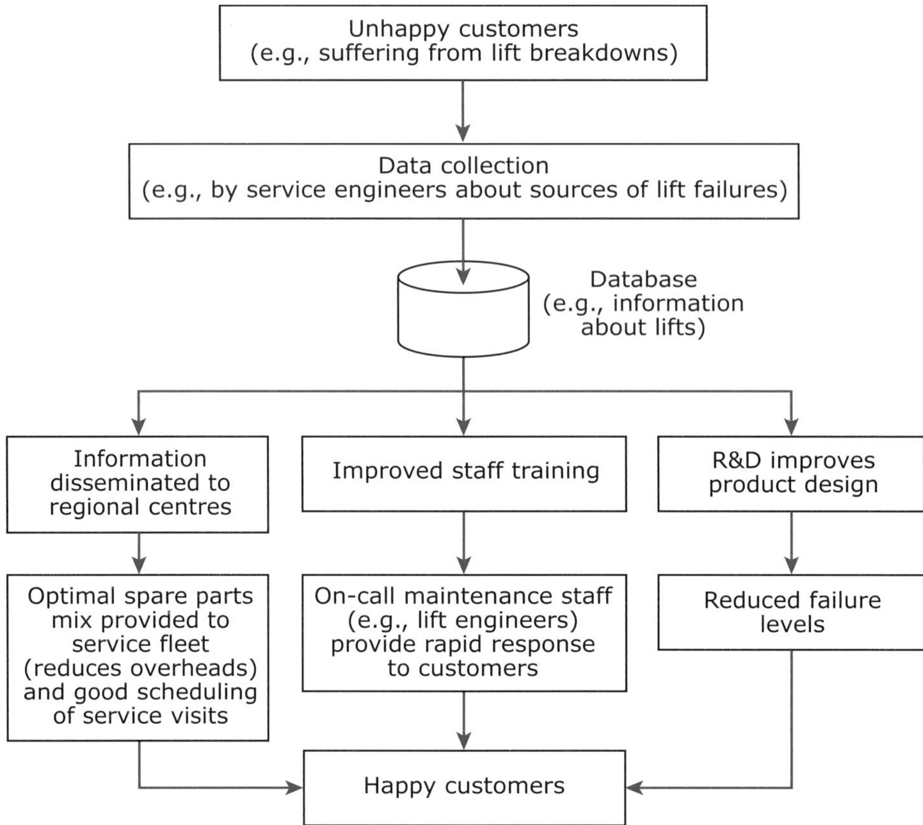

Figure 13 Benefits of effective database management: the Otis view

4.1.2 Enterprise resource planning (ERP) systems

Computerised ERP systems facilitate the integration of all the various functions of an organisation by interlinking information bases.

Many large organisations now use **Enterprise Resource Planning (ERP)** systems. At the heart of an ERP system, there is a central database that feeds data into modular applications relating to each function of the organisation: production, purchasing, accounting, marketing, human resources and so on. Use of an ERP system allows, for example, a direct linkage between receipt of a customer order, checking the customer's credit status, planning and scheduling production of the goods concerned, checking the inventory levels of the materials and components required, and placing purchase orders with suppliers for materials and components that need to be purchased.

This integration of information systems also facilitates planning and budgeting by enabling the implications of various scenarios to be explored. For example, the impact of various sales volume and sales mix projections on production capacity and manpower planning can be ascertained.

4.1.3 Advantages and disadvantages of database systems

The advantages of databases are:

- they avoid the unnecessary duplication of data
- as data are held only once, the possibility of different applications holding conflicting data on the same subject is eliminated
- data stored are independent of individual applications (programmes) that use the data, allowing greater flexibility in the way data can be used. Additional programmes can easily be introduced to make use of the existing data in a different way.

The disadvantages relate to issues of security and control:

- there is the possibility of unauthorised access to the data, necessitating procedures for data security and software controls
- as data are held only once but are used by many applications, the impact of systems failure is likely to be greater
- also, as data are held only once, the impact of inaccurate data is likely to be greater
- costs of setting up a database system can be very high.

Information system security

As you have seen, information is a precious resource for most organisations. It is essential, therefore, to ensure that data and information are reliable and accurate. It is also essential that data and information are kept secure. Security refers to the protection of data from unauthorised changes, uses or destruction. There are several measures that organisations can take to protect data and information, which may be classified as described here.

- *Security controls* protect data from unauthorised modification, disclosure or destruction and from non-availability of services to users.
- *Integrity controls* maintain the correctness and completeness of data, for example, ensuring that data stored electronically are the same as in the source documents.
- *Contingency controls* are for dealing with unscheduled interruptions to information systems processing. These are sometimes referred to as **disaster recovery plans**.

4.1.4 The use of multimedia in database applications

The content of databases is no longer restricted to words and numbers. Database applications can now include storage and use of multimedia, such as photographs and scanned images of documents. The use of multimedia in database applications has become widespread as organisations take advantage of the capabilities of storing large amounts of graphical and textual information. The advent of the Internet, both as a medium for marketing and electronic commerce, and as a model for **intranets** and **extranets** (an intranet extended beyond the boundaries of the organisation, with secure protected access), has led to increasing demand for images, audio and video material in databases.

Multi-media databases

Databases of insurance policies use multi-media. They store photographs of damaged property, handwritten claim forms, audio transcripts of loss appraisals, images of insured objects and even video walk-through of houses. Image files are large and research has delivered increasingly powerful and sophisticated compression software to reduce the size of stored versions of such image files.

Estate agents are using multi-media for selling houses. You can view as many houses as you wish on your PC and then pick those you want to visit, taking with you the property visit schedule without ever having gone near the estate agent. Holiday companies and estate agents are placing property and hotel information on multimedia to increase value to customers and clients. For prestige items, there are virtual walk-throughs of cars or houses. Companies use multi-media and the Internet to make the details of their **annual general meetings (AGMs)** available to all shareholders, not just those who can travel to where they are being held.

4.1.5 Data warehousing and data mining

Databases now have the potential of being so large – becoming vast **data warehouses** – that no one could ever be sure of finding all relevant information within them without the aid of some specialist software. This has obvious negative implications for data management. It also has huge potential benefits if a means is found to tap into these data warehouses and release the information held within them. This situation has led to the development of **data mining**.

Data mining simply involves the computer searching and analysing the data for you. Somebody once said that it is like mining for gold with a spoon. The computer will work away and find you multitudes of relationships, but which ones do you act upon? The commercial decision is yours.

Typically, data warehouses:

- are used by large organisations
- contain high volumes of data
- utilise data from transactions systems
- are driven by information needs.

The use of data warehousing and data mining are appealing to organisations because of the vast amounts of data it is possible to capture. Retail organisations are some of the major users of the techniques because of the opportunity provided to capture data about purchasing patterns at the point of sale in shops. However, any such extensive investment in data storage, handling and analysis needs to be ongoing. All investment and development cannot cease as soon as the data warehouse is 'up and running'.

Fortunately, while the costs of managing and using data may be increasing, the cost of data capture is being reduced by making additional use of data captured for one purpose, typically at point of sale in shops or through order forms. This flexibility is not as great in practice as it could be in theory. Data protection legislation places significant duties on the keepers of such information and increasing restrictions on its use for anything other than its original purpose.

4.2 Information markets and the Internet

The term **information market** reflects the view that information is a commodity that can be bought, sold or exchanged according to terms and conditions agreed between buyer and seller.

An important theme of this unit is the benefits which information, properly managed and used, can bring to an organisation. The way in which information markets have emerged is best illustrated by the phenomenal growth of the Internet.

4.2.1 The Internet

The Internet is the name given to the global information network that allows any computer with a telecommunications link to send and receive information from any other suitably equipped computer. The Internet has existed for a great many years, but it was only with the development of the **graphical user interface** for the **browser software** in the early 1990s that it developed its mass appeal. Even then, it took the best part of a decade for it to become what it is today: an integral part of business, commerce, and everyday life.

Most companies of any size now have a website – a collection of screens (web pages) providing information in text and graphic form, any of which can be viewed simply by clicking the appropriate button, word or image on the screen. Many of these websites are simply an additional outlet for information about an organisation, rather than full-blown attempts at marketing a product or a service. People looking at web pages do not tend to return often to pages that are not regularly updated. Therefore, where information is made available through this medium, it needs to be effectively managed and kept visibly up to date.

'I don't know what plagiarising is, so I'm gonna take the easy way out and just copy something off the Internet.'

Any information that can be presented on paper, through video or by sound can be presented on a web page. In addition, web pages can have interactive flexibility enabling real time graphics to be portrayed on request: see, for example, web pages showing share price movements. Even more information can be provided to the visitor to a website through the use of virtual reality software. This can, for example, enable prospective customers to see products at any angle and tourists to view in detail the sights they will see personally should they choose to visit them.

4.2.2 Commercial use of the Internet

Marketing

In some industries, the Internet now enables customers direct access to product creation. Customers can directly establish their personal preferences concerning product design.

Besides using it to tap into worldwide information resources, businesses are also using the web to provide information about their products and services. Companies identify and target particular markets and customers. The Internet offers a speedy and impersonal way of providing the basics (or even the details) of the services that a company offers. It offers a cost advantage over staffed enquiry desks or mail shots (that people may throw away or forget about) and is available when the information is needed. However, some companies have invested considerable resources in elaborate websites, with little evidence of any direct return on that investment.

Business use of the Internet has grown for products or services about which customers are liable to engage in information seeking, either for intermittent, relatively high value purchases such as washing machines, TVs, motor cars, holidays, etc., or for specialist items, such as out of print books.

Sales

Interactive electronic purchasing is now common, although there is still some resistance among potential users. People are reluctant to supply their credit card details to a computer, although they will very gladly do so in a one to one transaction in a shop or over the telephone, and happily transcribe the details on to a bill that they are paying by post. Nevertheless, customers need to be reassured about security and websites which offer online sales generally use secure software to transfer the transaction data.

Further technological developments will provide whatever reassurance is needed and, once people are used to this method of buying, it is likely to become the norm for many transactions that are presently conducted by post, fax or telephone.

Distribution

The Internet can be used to get certain products directly into people's homes. Anything that can be converted into digital form can simply be uploaded onto the seller's site and then downloaded onto the customer's PC at home. The Internet thus offers huge opportunities to producers of text, graphics/video, and sound-based products. Much computer software is now distributed in this way.

4.2.3 Internal communication: intranets

An *intranet* is a network using *Internet* technology within an organisation to support its work that is accessible only within that organisation.

The idea behind an intranet is that companies set up their own small-scale version of the Internet, using a combination of the company's own networked computers and Internet technology. Each employee has a browser and a server computer distributes corporate information on a wide variety of topics, and also offers access to the global Internet.

Potential applications include daily company newspapers, induction material, an online procedures and policies manual, employee web pages where individuals post details of their activities and progress, and internal databases of the corporate information store, including current and up to date financial information.

4.2.4 Telecommunications dangers

When data are transmitted over a telecommunications link (especially the Internet) there are numerous security dangers.

Activity 4.3

We have already considered the obvious concerns of online customers over potential fraud.

Can you think of additional security concerns that a company considering the development of an intranet might have?

Feedback

You may have come up with a number of the following, some of which are general to all networked computer systems carrying sensitive information.

- Corruption, such as viruses on a single computer, can spread through the network to all of the organisation's computers.
- Staff can do damage through their own computer to data stored on other computers. For example, they may try to transfer a file from their own hard disk to a colleague's hard disk without realising that their colleague has a file with the same name. Unless care is exercised it is easy to overwrite somebody else's data.
- Disaffected employees have much greater potential to do deliberate damage to valuable corporate data or systems because there is an increased risk that they may gain access to parts of the system to which they should not have access.
- If the organisation is linked to an external network, persons outside the company (hackers) may be able to get into the company's internal network, either to steal data or to damage the system.
- Employees may download inaccurate information or imperfect or virus-ridden software from an external network. For example, 'beta' (free trial) versions of forthcoming new editions of many major packages are often available on the Internet, but the whole point about a beta version is that it is not fully tested and may contain 'bugs' that could disrupt an entire system.
- Information transmitted from one part of an organisation to another may be intercepted. Data can be 'encrypted' (scrambled) in an attempt to make it unintelligible to electronic eavesdroppers.
- The communications link itself may break down or distort data. The worldwide telecommunications infrastructure is improving owing to the use of new technologies, and there are communications 'protocols' governing the format of data and signals transferred. At present, however, transmitted data are only as secure as the medium through which they are transmitted.

Although utilising databases and technology allows organisations to handle large amounts of data and provides opportunities for using information intelligently, it has created many problems. Data warehousing and data mining are possible, but there is still the need for the manager to make the decisions about the value of the information identified.

The Internet also opens up tremendous possibilities for using different methods of selling, but technology takes time to become accepted by the user and there may still be a degree of resistance to its widespread use for commercial transactions at an individual level. The glamour and sophistication of some of the applications means that management does not always apply the same rigour to investment decisions in these new forms of marketing and sales support. There is a need for accurate and realistic costings.

4.3 Knowledge work

Technology allows organisations to collect and analyse large amounts of data. However, it is still the manager who makes the final decision about what to collect in many cases, particularly in decisions where a high degree of uncertainty exists. **Knowledge work** is dependent upon the knowledge possessed by the worker, not on ability to perform physical tasks. Most tasks in an organisation include some element of knowledge work.

'What choice do we have? I can't fire him. He knows too much.'

4.3.1 Attributes of knowledge work

Key attributes of knowledge work are that:

- it involves thinking, processing information, and formulating analysis, recommendations and procedures
- it may use verbal or written inputs and outputs and multi-media.

In general, managerial jobs are more likely to contain a high level of knowledge work and clerical jobs are likely to contain a lower level of knowledge work. However, it would be a gross over-simplification to equate knowledge work with managerial work. Some managers may perform a high degree of routine work and therefore do little knowledge work. However, accounting tasks involve a lot of knowledge work. While bookkeeping is a routine and low level activity, **tacit knowledge** is essential to accounting tasks, such as interpreting financial ratios or selecting appropriate **discount factors** for discounted cash flow computations. (These management accounting techniques will be covered in later units of this module.)

Types of knowledge work include:

- *diagnosis and problem finding*, such as the identification of potential bad debts (irrecoverable receivables)
- *planning and decision making*, such as capital investment appraisal
- *monitoring and control*, such as the review of implemented capital investment projects

- *authoring and presentation*, where knowledge work progresses from an idea through multiple media to a final presentation, such as a tender presentation given to a potential customer for audit services
- *communication*, as a large part of the time of a knowledge worker is spent communicating, either obtaining information so as to gain knowledge, or passing on knowledge.

4.3.2 Information preference

As previously outlined, knowledge work typically involves information. It must, for if you remember the discussion in Session 3, knowledge is derived from information. However, managers have been found to prefer 'live' information. They:

- place greater reliance on up to date information even if it may not be thoroughly verified
- tend to favour verbal information, telephone conversations or formal and informal meetings.

As a consequence, much of the information upon which decisions are made is undocumented and qualitative. This tends to create resistance to decision making techniques that rely on amassing large amounts of quantitative information such as can be provided by computerised information systems.

When vital **qualitative information** is locked inside the head of a manager, this information can only be incorporated into a decision if that particular manager is actively engaged in the decision making process. No matter how sophisticated the techniques used, if a vital piece of information, whether qualitative or quantitative, is not available to the decision maker, the information provided is likely to be viewed differently, resulting in wrong decisions being taken. Decisions taken without the fullest knowledge of the issues are less likely to be correct than those taken where everything possible is known.

It is essential that such examples of the problems that can result from the existence of personally held information be overcome through effective knowledge management.

Activity 4.4

How does this preference for 'live' information affect the role of accountants, whose traditional role is based on a concern for verifiable and quantifiable information derived from 'hard' data (i.e., numbers and numerical analysis)?

Feedback

If managers prefer 'live' information to quantitative data and information, and if accountants base their role upon the provision of quantitative data and information, accountants are unlikely to provide managers with some of the key information they would like. When some people in an organisation have qualitative information that is only ever expressed if they are involved in the decision making process, accountants need not to only embrace qualitative data and information within the service they provide, but also to identify such individuals and involve them in those decisions.

4.4 Technology and decision making

Activity 4.5

How do you think technology can assist the decision making process?

Feedback

Technology can help with the decision making process as the information gathering capabilities and analysis tools available are more numerous and provide attributes that the human being cannot. However, technology will only help in certain situations. Computers are good at handling large amounts of quantitative data and, therefore, will be more helpful for the type of problem that can be solved using quantitative techniques than for a problem requiring a good deal of sensitive handling and qualitative analysis.

Most significant decisions involve risk or uncertainty. The information and knowledge used by decision makers and the decision making process itself need to recognise that things are not 'black and white', and that there is a degree of risk and uncertainty in most decision making situations.

Technology helps to gather more information and thus reduce the uncertainty, by allowing speedy analysis and the evaluation of different possible options

Programmed and non-programmed decisions

In terms of computer systems, it is appropriate to distinguish between programmed and non-programmed decisions. **Programmed decisions** are those that can be pre-specified by a set of rules or decision procedures and are thus eminently suitable for computers to handle. In such decision making situations, the relationship between ends and means is well understood and so detailed instructions may be provided on exactly how a task should be performed.

Non-programmed decisions have no pre-established decision rules or procedures. They are ill-structured, not repeated frequently, or conditions are so different at each repetition that no general model can be developed as a basis for programming them. **Decision support systems** are suitable for this type of decision. In such decisions, the relationship between ends and means is not well understood. Consequently, it is not possible to provide detailed instructions as to how a task should be performed and it is necessary to rely on the judgement of managers.

Decision making in accounting

The distinction between these two types of decision situations affects the type of accounting techniques that are appropriate for management control. For programmed decisions, the inputs required to produce a given output can be specified in advance with a high level of accuracy. For example, it is often possible to specify accurately the amount of labour and materials required to manufacture a product. Accountants can then specify detailed standards for inputs usage and cost and monitor actual usage and cost against these. Any significant deviation from standard can then be investigated and corrected by management. An example of this approach is the technique of standard costing and variance analysis, which you will look at in Unit 4 of this module.

For non-programmed decisions, it is not possible to specify clearly what the inputs required to produce a given output should be. Managers must use their judgement and managers are held accountable only for achieving the desired output. An example of such a situation arises in the operating divisions of large, diverse companies. Such divisions will typically be required to maximise profit or return on investment, but how they do this will depend on the intuition, skill, knowledge of markets and operations of divisional management. Divisional management will be measured according to results achieved. There will be no attempt to specify in advance the details of how the task should be performed and then monitor adherence to the prescribed details, as in the case of programmed decisions.

(Source: adapted from Emmanuel et al., 1990, pp. 15–6)

4.4.1 Expert systems

When decision making involves tasks that require expertise and specialised knowledge, but the tasks are performed so infrequently as to mean that decision makers may not ever develop the appropriate level of expertise, a special form of decision support system called an **expert system (ES)** can be used. This is particularly the case where expertise is rare and non-experts must carry out the task.

An ES is a computer application that attempts to emulate the output of an expert when solving a specific problem. Expert systems are also sometimes referred to as **knowledge-based systems**. In 2009, the UK government responded to a flu pandemic by setting up a telephone helpline where non-experts asked callers a set of questions designed by experts to decide whether the caller should be authorised to receive 'Tamiflu' medication. This is a good example of the application of an expert system.

A non-expert can perform at the level of an expert using this form of software. Examples in practice range from detecting credit card fraud to corporate lending decisions by banks. VAT registration and **accounting standards** are two areas for which expert systems have been very successfully developed. In recent years, expert systems have made a significant contribution to the practice of accounting. In auditing, expert systems are used for everything from audit planning through to identification of **going concern** problems. In financial accounting, expert systems are used to determine the financial status of an organisation by analysing its financial statements. In management accounting, expert systems are used to evaluate investment decisions, establish transfer prices for interdivisional trading, analyse deviations from budget for cost control, evaluate divisional performance and even design organisation-wide management accounting systems.

A number of potential problems is exhibited by the development of expert systems.

- They can be costly to develop. Some large experimental systems have required from 10 to 25 person years of effort and millions of pounds of investment.
- Eliciting knowledge from an 'expert' to incorporate into an expert system may be costly and difficult. One reason is that, as experts establish methods to aid their search for a solution, these methods are often not implemented in a programmable fashion. Another is that experts are not often aware of the extent of their own knowledge, nor can they express intelligibly to someone else what they do know.

- Expert systems may not be effective beyond the sets of examples and experiences of the experts used to build them. They may not, therefore, be reliable in novel situations.
- An effective expert system requires a simple but flexible user interface for the non-expert user. It needs to be user-friendly and not so ponderous as to reduce any value added by using the expert system.
- Users may not accept the output of an expert system if it is not transparent in its reasoning and decision making processes and many are not transparent, particularly those expert systems that deal with complex problems such as whether or not a company is likely to fail.

Activity 4.6

Spend a few minutes listing as many reasons as you can think of for users to reject an accounting expert system even if it is shown to be at least as good as a human expert.

Feedback

On the list you prepared, you may have included any or all of the following:

- resistance to technology
- uncertainty over the boundaries of the expert system's knowledge base
- blind belief that human beings are always superior to machines
- lack of adequate explanations from the expert system concerning why it has recommended a particular course of action
- belief by the individual that they are more of an expert than the expert system
- misalignment of responsibility for the decision being taken, that is, if the user is responsible for the decision even though it was recommended by the expert system.

4.4.2 Advantages of using expert systems

The advantages of expert systems include the following.

- *Provision of cost savings*, in that they are capable of evaluating many more possible options within a given time frame than is possible for a human brain.
- *Lack of emotion*, meaning that the decision will be made based on the facts alone.
- Having better *memory capability* than humans which can be designed to remember the previous decisions taken based on similar facts. When designed in this way, they can demonstrate an ability to learn from previous decisions by virtue of their ability to remember.
- They can discover rules in sets of expert judgements and in effect convert tacit knowledge into explicit rule-based knowledge.
- They capture the knowledge of experts, who are retiring, so that knowledge is retained within the organisation.
- They are potentially available 24 hours a day, 365 days a year.
- They are consistent, that is, they always give the same recommendation when the circumstances are identical.

Expert systems involve various forms of knowledge representation. The term **knowledge base** is used in place of database to indicate this. Once computers are applied to areas involving knowledge and judgment, they can aid organisations significantly.

Summary

This session has provided an introduction to some important ways in which information and communication technologies can be exploited to enhance the capture, storage and distribution of information in an organisation. It has also considered the nature of knowledge work and the use of technology for decision making.

Unit summary

You have now completed Unit 1, the first unit of B292 *Management accounting*. This unit has acted as an overall introduction to the module by considering the context of management accounting. A number of concepts and ideas to which you have been introduced in this unit will be explored and developed in subsequent units.

By now, you should have an understanding of the nature of organisations, the management processes of planning and control, the role of information (including accounting information) in managing organisations and the opportunities presented by information technology for capturing, storing and distributing information in an organisation. This all provides an excellent foundation for developing the specific subject matter addressed in subsequent units.

Before moving on to Unit 2 of this module, attempt the following self-assessed questions. Make sure you check your answers with those provided before you start Unit 2.

Self-assessed Questions

(Where appropriate use examples to support your arguments.)

Question 1

Why do you think management accountants need a good understanding of the nature of organisations (including their structure and objectives) and the environment in which they operate?

Suggested answer

Management accountants are an important part of the management team, providing decision support and control information to the managers of the various functional areas/departments of the organisation. Increasingly, management accountants are acting as 'business partners' with other functional areas in the organisation rather than remaining in their 'silos' in the central finance team. In order to provide customised and relevant information, management accountants must understand the technology, operations and management structure of their organisation.

Management accountants are producing information which is intended to help managers make decisions about the organisation, so understanding the nature of the organisation and its objectives is important in determining what information managers will require. Often, functional area managers will not know exactly what accounting information and analysis could be made available to them, so a continuing dialogue between accountants and other managers is necessary. Often accountants will help other managers interpret the information provided to them in such a way as is relevant to their needs. A meaningful dialogue can only happen if accountants have an understanding of the functional area concerned in terms of its objectives and operations. Accountants must also understand the formal structure of the organisation in order to design and operate an appropriate system of responsibility accounting, which provides essential information for management control.

Organisations operate in a wider economic environment. Changes in this environment will impact on the organisation, for example, interest rate movements, anticipated inflation, government tax policy, projections of economic growth (or decline). Financial planning and budgeting clearly require the systematic monitoring and collating of information about these factors. Such will usually be the responsibility of the organisation's Accounting and Finance function.

Question 2

From your observations of managers, compile your own list of the most important functions a manager has to perform. You can use any management role as a prompt and it need not be a manager in a business environment. You might think, perhaps, of someone who manages a household.

Suggested answer

Your list of management activities might include the following:

- planning
- organising
- securing resources
- making decisions
- handling information
- communicating
- negotiating
- meeting deadlines and performance targets
- thinking creatively
- managing change situations
- delegating
- dealing with internal politics
- developing and motivating staff
- resolving conflicts
- dealing with problems and crises
- forecasting
- networking.

This list is not exhaustive and your list may be slightly different. This list contains some of the functions that Fayol described – planning and organising in particular. There are also elements of coordination and control in the sense that managers have to ensure the smooth functioning of the part of the organisation for which they are responsible in order to meet targets.

This list implies a very different inter-relationship between the manager and staff and the manager and the organisation as a whole. For example, the manager today is less likely to command and more likely to negotiate, participate and motivate. He or she will spend a good deal of time reacting to or anticipating changes in the environment, hence the need for creative thinking and decision making.

Question 3

Research suggests that practising managers often do not use accounting information in their planning, control and decision making activities. Why do you think this is so?

Suggested answer

Managers may not think it is relevant, perhaps because they do not understand accounting information (if it is quantitative, it may be off-putting) and the language accountants use. They may associate it with being controlled and measured and not want accounting information used in the decision, fearing that after the decision, such information will be used to measure their performance. They may distrust accountants, seeing their role as 'policing' the activities of others.

It has also been suggested (e.g., Johnson and Kaplan, 1987) that it is because accounting information is often provided too late to be of use in control or decision making activities and is not in an appropriate form to be of use. For example, accounting information is often in the form of highly aggregated financial data, whereas managers require disaggregated, detailed non-financial data to take the appropriate actions.

Another theory is that managers often do not use accounting information because it is not relevant to their (as opposed to the organisation's) personal needs. Managers' objectives (it is argued) are often not those assumed by the providers of accounting information. In particular, managers have their own personal objectives and agendas and are not necessarily primarily concerned with achieving the organisation's objectives. Most management accounting systems, on the other hand, are designed to provide information to enable managers to achieve organisational objectives, for example, maximising profits.

Question 4

Emmanuel et al. (1990, p. 107) have suggested that as a manager rises in the organisational hierarchy, he/she knows less and less about more and more. Explain why you agree or disagree with this statement.

Suggested answer

At higher levels in the organisational hierarchy, managers typically have a wider area to manage and as a result can have less detailed knowledge about individual operations and functions than those lower in the hierarchy who are 'closer to the action'. Indeed, this is one of the main reasons for the emergence of a hierarchy in organisations. As organisations grow in size and complexity, it becomes impossible for senior managers to know the details of all sections of the organisation and delegation of decision making authority to those with more detailed and specialised knowledge is necessary. A Chief Executive Officer (CEO) knows the whole organisation, but not in the same level of detail that the manager of a department knows about his/her department.

Question 5

This unit has looked at the management of data and information through database and Internet technologies. Identify the barriers to customer and user acceptance of exchanging Internet informaion.

Suggested answer

There is a number of barriers to user acceptance of information exchange via the Internet. First users may not trust the source of the information. The viewer of a web page containing, for example, the financial statements of an organisation (the income statement, balance sheet and so on) may be unaware whether or not the information has been verified before being placed on the web. Even when it has apparently been verified, how does the viewer know for sure that the verification is genuine?

There is also the problem of possible lack of understanding of how to access the relevant information. An additional possible problem is being presented with too much information, that is, the problem of 'information overload'.

Question 6

This unit has introduced some key types of information system used to support decision making. Why do you think it is necessary to move beyond quantifiable 'hard' data in order to make robust decisions in organisations?

Suggested answer

Quantifiable data does not give the complete picture. In order to make good decisions, it is necessary also to consider the qualitative or 'soft' data relevant to a situation. For example, when a senior manager is evaluating the performance of a subordinate, say a profit centre manager, he will have the numbers concerning actual and budgeted profit performance. In order to decide, however, whether the subordinate manager's performance has been satisfactory, in the event, say, of failing to achieve the budgeted profit level, the senior manager will need to consider the actual circumstances faced by the subordinate manager. Perhaps the profit centre manager was faced with the prolonged absence, due to illness, of key personnel or a change in economic climate.

In addition, it is often not possible to quantify all the objectives and achievements of a subordinate and senior managers must rely on trust rather than accounting numbers for control. Consider, for example, the CEO evaluating the managers of the Human Resources or Finance functions!

References

Table of statutes

Great Britain. *Companies Act 2006. Elizabeth II. Chapter 46.* (2006) London: The Stationery Office.

Great Britain. *Consumer Credit Act 1974. Elizabeth II. Chapter 39.* (1974) London: The Stationery Office.

Great Britain. *Consumer Credit Act 2006. Elizabeth II. Chapter 14.* (2006) London: The Stationery Office.

Great Britain. *Data Protection Act 1998. Elizabeth II. Chapter 29.* (1998) London: The Stationery Office.

Great Britain. *Sale of Goods Act 1979. Elizabeth II. Chapter 54.* (1979) London: The Stationery Office.

Publications

Anthony, R.N. (1965) *Planning and Control Systems: A Framework for Analysis*, Boston, MA, Division of Research, Harvard Graduate School of Business.

Bentley, T.J. (1998) *Managing Information: Avoiding Overload*, London, Chartered Institute of Management Accountants, Kogan Page.

Chandler, A.D. (1962) *Strategy and Structure*, Boston, MA, MIT Press.

Coates, J.B., Rickwood, C. and Stacey, R.J. (1996) *Management Accounting for Strategic and Operational Control*, Oxford, Butterworth-Heinemann.

Davenport, T.H. and Prusak, L. (1997) *Working Knowledge*, Boston, MA, Harvard Business School Press.

Emmanuel, C., Otley, D. and Merchant, K. (1990) *Accounting for Management Control* (2nd edn), London, Chapman and Hall.

Fayol, H. (1949) *General and Industrial Management*, London, Pitman.

Harrison, G.L. (1978) 'The role of the management accountant in the marketing function of business: case observations and analyses', Research paper no. 157, School of Economics and Financial Studies, Macquarie University.

Johnson, H.T. and Kaplan, R.S. (1987) *Relevance Lost: The Rise and Fall of Management Accounting*, Boston, MA, Harvard Business School Press.

Kotler, P. (1967) *Marketing Management*, Englewood Cliffs, NJ, Prentice Hall.

Mintzberg, H. (1978) 'Patterns in Strategy Formation', *Management Science*, vol. 24, no. 9, pp. 934–48.

Mintzberg, H. (1979) *The Structuring of Organisations*, Englewood Cliffs, NJ, Prentice Hall.

Porter, M.E. (1985) *Competitive Advantage: Creating and Sustaining Superior Performance*, New York, Free Press.

Rockart, J.F. (1979) 'Chief executives define their own data needs', *Harvard Business Review*, vol. 57, no. 2, pp. 238–41.

Acknowledgements

Grateful acknowledgement is made to the following sources:

Figures

Figure 1: Mintzberg, H. (1979) 'Five basic parts of organisations', The Structure of Organizations. Pearson Education, Inc.

Figure 10: Ellis, J. and Williams, D. (1993) 'Assessing strategic capability', Corporate Strategy and Financial Analysis, Pitman Publishing.

Illustrations

Page 9: With kind permission from Philip Talbot

Page 13: With kind permission from Philip Talbot

Page 19: With kind permission from Philip Talbot

Page 25: © John Morris, www.CartoonStock.com

Page 27: © Mike Shapiro, www.CartoonStock.com

Page 33: © Mike Shapiro, www.CartoonStock.com

Page 38: Reproduced with permission from The Spectator (1828) Ltd.

Page 45: With kind permission from Philip Talbot

Page 54: © Aaron Bacall, www.CartoonStock.com

Page 60: © Andrew Toos, www.CartoonStock.com

Page 63: © Adey Bryant, www.CartoonStock.com

Page 67: © Mike Flanagan, www.CartoonStock.com

Page 73: © Marty Bucella, www.CartoonStock.com

Page 76: © Mike Baldwin, www.CartoonStock.com

Every effort has been made to contact copyright holders. If any have been inadvertently overlooked the publishers will be pleased to make the necessary arrangements at the first opportunity.